MONTESCANO (PAVIA) ITALY

Current Topics in Rehabilitation
Series Editor: R. Corsico M.D.

Forthcoming titles in the series:

Pulmonary Circulation
Edited by: F. Cobelli - M. Morpurgo - C. Fracchia - C. Rampulla - R. Tramarin

Pulmonary Hyperinflation
Edited by: A. Grassino - C. Rampulla - N. Ambrosino - C. Fracchia - L. Zocchi

Acknowledgments: the Organizers of the Workshop "Respiratory Muscles in C.O.P.D.", held in Montescano (Pavia) on September 8-9 1986, are indebted to Camillo Corvi S.p.A. (Piacenza) for their support in publishing this volume.
Thanks to Peter Mead, M.A. (Oxon), for the linguistic revision of the text.

Respiratory Muscles in Chronic Obstructive Pulmonary Disease

Edited by: A. Grassino, C. Fracchia, C. Rampulla, L. Zocchi

Springer-Verlag Berlin Heidelberg GmbH

A. Grassino – Meakins-Christie Laboratories, McGill University and Notre-Dame Hospital, University of Montreal, Quebec, Canada
C. Fracchia – Medical Centre of Rehabilitation, Montescano, Pavia, Italy
C. Rampulla – Medical Centre of Rehabilitation, Montescano, Pavia, Italy
L. Zocchi – Meakins-Christie Laboratories, McGill University, Montreal, Quebec, Canada

Series Editor:

R. Corsico – Medical Centre of Rehabilitation, Montescano, Pavia, Italy.

Italy:
BI & GI EDITORI - Via Cantarane, 6 - P.O. Box 2106 Verona 2 - 37100 Verona, Italy

North America:
SPRINGER - VERLAG New York, Inc. - 175 Fifth Avenue, New York, N.Y. 10010, USA

Japan:
SPRINGER - VERLAG - 37-3 Hongo 3-chome, Bunkyo-Ku, Tokyo 113, Japan

Rest of the World:
SPRINGER - VERLAG - Berlin Heidelberger Platz 3, 1000 Berlin 33, FRG

ISBN 978-1-4471-3852-5

Library of Congress Cataloging-in-Publication Data
Workshop "Respiratory Muscles in C.O.P.D." (1986: Montescano, Italy)
Respiratory Muscles in Chronic Obstructive Pulmonary Disease. (Current topics in rehabilitation)
"Workshop "Respiratory Muscles in C.O.P.D.", held in Montescano (Pavia) on September 8-9, 1986"--Ser, t.p.
1. Lungs-Diseases, Obstructive-Pathophysiology-Congresses. 2. Respiratory muscles-Congresses. 3. Lungs-Diseases, Obstructive-Patients-Rehabilitation-Congresses. I. Grassino, A. II, Title III, Series [DNLM: 1. Lung Diseases, Obstructive-physiopathology-congresses. 2. Respiratory Muscles-physiopathology-congresses. WF 600 W9255r 1986] RC776, 03W67 1986 616.2'4
88-2008
ISBN 978-1-4471-3852-5 ISBN 978-1-4471-3850-1 (eBook)
DOI 10.1007/978-1-4471-3850-1

Originally published by Springer-Verlag Berlin Heidelberg New York - Tokyo in 1988
Softcover reprint of the hardcover 1st edition 1988

Typeset by Composit, Verona - Italy
Printed by Gutenberg, Povegliano - Verona - Italy

Preface

While emphysema and chronic bronchitis are primarily lung diseases, one of their major consequences is to deeply affect the function of the respiratory muscles. Lung hyperinflation shortens the inspiratory muscles due to increased airways resistance, more of their effort is demanded and changes in nutritional status weaken them further. Their malfunction can lead to severe dyspnea and to failure of the ventilatory pump. Over the last 10 years we have witnessed an explosion of information of how respiratory muscles function in health and disease, new techniques for their evaluation have been created, the concept of fatigue, weakness, and failure was developed, and their rest or training was attempted. The implication of respiratory muscles malfunction in respiratory medicine has reached a prominent place.

It seems remarkable that while some aspect of skeletal muscles function requires molecular biology techniques to find new answers, we still know little on respiratory muscles interaction, strategies of coordination, their role in dyspnea, chronic hypercapnia or how to effectively improve their function in patients.

This workshop was organized and held at the Medical Center of Rehabilitation in Montescano and represents an attempt to focus on how the newly adquired wealth of information can eventually be trasformed into medical care.

The participants in this workshop brought forward challenging thoughts and we are most grateful for their participation. This book represents a report of the proceedings and also provides the most updated information in this field.

Alex Grassino

Foreword

Research centres specializing in rehabilitation medicine must keep abreast of the latest scientific developments in physiopathology. Only by constant striving in this direction can functional impairment be efficiently evaluated and recovery be correctly monitored.

Rehabilitation often calls for a multidisciplinary approach. The scientific meetings periodically organised at the Rehabilitation Centre at Montescano set out to meet this need, providing a variety of approaches to specific areas of interest.

"Current Topics in Rehabilitation" is a series based on the items presented and discussed at the Montescano workshops. It is, then, fitting that this short preface should offer a work of thanks to the contributors without whose skill and commitment these meetings could not be held.

In this series, our aim is to offer the reader an insight into the discussions and conclusions of each workshop, as well as giving due recognition to the valuable scientific work of the Montescano Rehabilitation Centre.

It is our wish to dedicate the series to the memory of Prof. Salvatore Maugeri, the distinguished medical researcher who founded our Centre and to whose foresight we are grateful for clarifying the aims it is our privilege to pursue.

Renato Corsico
Centre of Rehabilitation Medical
Montescano (Pavia), I.R.C.C.S.

November 1987

Contents

Contributors

ARRIGO, A.
Chair of Clinical Neurophysiology, University of Pavia, "C. Mondino" Foundation, Pavia, Italy.

BELLEMARE, F.
Meakins-Christie Laboratories, McGill University, Montreal, Quebec, Canada.

BELLIA, V.
Respiratory Physiopathology Department. National Research Council, Palermo, Italy.

BONSIGNORE, G.
Respiratory Physiopathology Department. National Research Council, Palermo, Italy.

BRONDOLESE, R.
Department of Anesthesia and Intensive Care, City Hospital, Padua, Italy.

BROSEGHINI, C.
Institute of Occupational Health, University of Padua, Italy.

BUONOCORE, M.
Clinical Neurophysiology Department, Medical Rehabilitation Centre, "Clinica del Lavoro" Foundation, Montescano, Pavia, Italy.

CASALE, R.
Clinical Neurophysiology Department, Medical Rehabilitation Centre, "Clinica del Lavoro" Foundation, Montescano, Pavia, Italy.

COLLET, P.W.
Laboratory of Experimental Medicine, Faculty of Medicine, University of Aix-Marseille, France.

D'ANGELO, E.
Institute of Human Physiology, University of Milan, Italy.

DEL CANALE, S.
Institute of Clinical Medicine and Nephrology, University of Parma, Italy.

DERENNE, J.P.
Pneumology Department, Hôpital Saint Antoine, Paris, France.

ENGEL, L.A.
Thoracic Medicine Unit, Westmead Hospital, Sydney, Australia.

ESTENNE, M.
Chest Service, Erasme University Hospital, Brussels, Belgium.

FIACCADORI, E.
Institute of Clinical Medicine and Nephrology, University of Parma, Italy.

FIELD, S.
Meakins-Christie Laboratories, McGill University, and Notre-Dame Hospital, University of Montreal, Quebec, Canada.

FITTING, J.W.
Division de Pneumologie, Départment de Médecine Interne, Centre Hospitalier Universitaire Vaudois, Lausanne, Switzerland.

GOTTFRIED, S.B.
Meakins-Christie Laboratories, McGill University, Montreal, Quebec, Canada.

GRASSINO, A.
Meakins-Christie Laboratories, McGill University and Notre-Dame Hospital, University of Montreal, Quebec, Canada.

GREEN, M.
Brompton Hospital, London, U.K.

GUARIGLIA, A.
Institute of Clinical Medicine and Nephrology, University of Parma, Italy.

JAMMES, Y.
Laboratoire de Physiologie Hyperbare, GS 15-CNRS, Faculté de Médecine Nord, Marseille, France.

MACKLEM, P.T.
Meakins-Christie Laboratories, McGill University, Montreal, Quebec, Canada.

MANZIN, E.
Department of Anesthesia and Intensive Care, City Hospital, Padua, Italy.

MIER, A.K.
Brompton Hospital, London, U.K.

MILIC-EMILI, J.
Meakins-Christie Laboratories, McGill University, Montreal, Quebec, Canada.

MOXHAM, J.
Brompton Hospital, London, U.K.

PARDY, R.L.
UBC Pulmonary Research Laboratory St. Paul's Hospital, Vancouver (B.C.), Canada.

PARIENTE, R.
Clinical evaluation of respiratory muscle fatigue, Inserm U 226, Hôpital Beaujon, Clichy, France.

POGGI, R.
Institute of Occupational Health, University of Padua, Italy.

ROMANO, S.
Respiratory Physiopathology Department, National Research Council, Palermo, Italy.

ROSSI, A.
Meakins-Christie Laboratories, McGill University, Montreal, Quebec, Canada.
Institute of Occupational Health, University of Padua, Italy.

SANCI, S.
Respiratory Physiopathology Department, National Research Council, Palermo, Italy.

SCHIAVINA, M.
S. Orsola Hospital, Bologna, Italy.

SMITH, J.
Meakins-Christie Laboratories, McGill University, Montreal, Quebec, Canada.

Aspects of Respiratory Muscles Physiology

A. Grassino - C. Fracchia - C. Rampulla - L. Zocchi
Respiratory Muscles in COPD
Bi & Gi Publishers, Verona - Springer-Verlag, London, 1988

Some old concepts in respiratory muscle mechanics

Jean Philippe Derenne
Pneumology Department, Hôpital Saint Antoine, Paris, France

SUMMARY: The article attempts a brief review of the major historical contributions to the study of respiratory muscle mechanics.
For all the originality of Aristotle's observations, they are of interest more as examples of physiology. After Galen's remarkable experiment some half a millennium later, there is an unfortunate gap in our knowledge of the great Abbasid isolated examples of Vesalius, Borelli and others, there are no further significant contributions untile the great French Experimenters of the 19th century.
Respiratory physiology subsequently suffered a neglect from which it has, even today, not fully recovered.

KEY WORDS: Respiratory muscles mechanics; Aristotele; Galen; Abbasid; Vesalius; Borelli; Respiratory physiology.

Respiratory muscle mechanics is one of the few areas of medical knowledge where the main themes have remained very little changed in the last centuries. If Aristotle or Galen came back today they would have difficulties in understanding the evolution of most of the fields of medicine. It is likely however that they would find our understanding of diaphragmatic function more familiar. Though we are still investigating the structure and function of the muscles and constantly introducing technical improvements, the basic concepts are essentially the same.

One of the most remarkable features of this field is that the discoveries made by the foremost scientists such as Galen, Vesalius, Magendie, Duchenne de Boulogne or Paul Bert have been forgotten by subsequent investigators. Furthermore the contributions of some major schools of medicine like the Abbasid or Cordoban schools have been wiped out from our memories, although their manuscripts can still be consulted.

In our culture, as everybody knows, the first scientific contributions in the fields in question, were made by the Greeks. The first was Aristotle (who lived in the 4th century BC). Because of the major importance of this author everybody knows his name, but few people are aware that he was the father of comparative zoology and that nearly 30% of his extensive writings are concerned with biology. His most important contributions in this field include the Peri Zoion Morion (De partibus animalium) and the Ton Peri ta Zoia Istorion (Historia animalium).

The former is the more important to us. It had a strong influence on his followers, particularly Plinius the elder and Galen, who plagiarized it thoroughly. It contains generalities on methods in biology and a description of the principles of biology based on the balance between opposite factors such as wetness and dryness, heat and cold, etc. Aristotle considered that the main function of respiration was to cool down the body. This conception was in fact unanimously accepted at that time and it seems that the first author to mention it was Empedocles.

In the Aristotelian conception, "the lung is the organ of respiration. It receives the principle of movement from the heart (i.e. it is the heart which is the main respiratory muscle) and offers a wide place to the incoming air because of its spongy nature and because of its size. In fact when it dilates, the air comes in and when it contracts, the air goes away" (Peri Zoion Morion III, 6, 669 a). The diaphragm isolates the heart and lungs from the abdominal cavity "in such a way that the main organ of the sensitive soul (the heart) does not feel any harm and is not hit by the inhalations released by digestion or by exhaled excess foreign heat" (III, 10, 672 a).

The diaphragm separates the top ("the better") from the bottom ("which exists for the benefit of the top"). It is "a kind of dam and a kind of fence".

Contrary to what is often said, Aristotle did not believe that the diaphragm was the center of the soul or of thought. He stated "When, in fact, because of the vicinity, these parts (the belly) attract hot and excremental humour, the latter brings immediately an obvious trouble to thinking and sensitivity, and that is why those parts are called phrenic as though they participated in thought.

"In fact that is not the case, but since they are close to those parts which participate, they make the changes which occur in reasoning obvious" (III, 10, 672 b). He gave an example of the influence of the diaphragm on mood, stating "war wounds in the region of the diaphragm provoke laughter because of the heat which leaks from the injury" (III, 10, 673 a). In fact the "sardonic smile", a peculiar type of smile, was related to diaphragmatic activity up to the 18th century AD.

Aristotle was a philosopher, not a physiologist. The first observations and experimental set-up of which we know now are those of Galen (end of the second century AD), who made a very precise description of the anatomy of the diaphragm (Adm. Anat. VIII, 1) and performed many experiments on dogs.

At that time most people thought that the diaphragm was the only muscle of inspiration. Galen discovered the cervical roots of the phrenic nerve and showed that transsection of the cord at the level of C7 did not harm the diaphragm, whereas transsection at C3 stopped any movement. He cut all the nerves supplying the intercostals, serratus, pectoralis and other muscles, and then he also cut off these muscles. In so doing he found that the diaphragm was inflating the lower rib cage.

When the diaphragm was paralysed and when the other muscles were intact, he found a paradoxical movement of the lower rib cage. Galen also described the cross-shaped intersection of the intercostal muscles and stated that the external were inspiratory and the internal were expiratory.

We do not know if these remarkable observations and experiments (which obviously show that Galen was a real respiratory muscle physiologist) were accepted or rejected in the 2nd century AD. We do not know, either, what the major chest specialists of the Middle Ages, i.e. Avenzoar, Avicenne or Averroes, all members of the rationalist scientific Arabic school, thought of this work.

Actually we know nothing of what they wrote. Nevertheless, there must be hundreds of manuscripts in various libraries, some related to respiration. Whether they are concerned with our field is something we would like to know. In fact, many scientists believe that the nature of the instruments their medieval predecessors used, as well as their surgical techniques, indicate that they performed human autopsies.

In western history, Vesalius (1543) and his followers were the first to perform autopsies. The father of anatomy was interested in the diaphragm and, after Galen, he wrote that "it pushes out the 6th and 7th ribs and... dilates the thorax and increases its capacity". Nevertheless, he strangely stated that "the diaphragm, when contracting during inspiration, was going up in the chest wall, and consequently was pushing the ribs upwards and outwards".

His fellow, Columbus, also professed that the diaphragm inflated the lower ribs but went downwards during inspiration. Nevertheless, he insisted that, in so doing, it was relaxed (during inspiration) and that, when contracting during expiration, moved upwards in the chest.

It seems that the first author to state that the diaphragm contracts and goes down during inspiration was Borelli. Nevertheless, he also stated that it deflated the lower rib cage and that the intercostals had no action on respiration.

After Borelli, major physiologists (Boerhaave, Haller) and anatomists (Winslow) of the 18th century stated the same beliefs.

The role of the intercostal muscles had been studied by many authors. It is impossible to summarize their works here and their conclusions were extremely variable. Amongst them, the most important was Hamberger, who made a physical model, showing that the external intercostals were inspiratory and the internal intercostals were expiratory.

In fact there were very few physiological hypotheses and investigations, at least in the modern sense of the term, before the 19th century. The first important contributor was Magendie (1816). Amongst his most important findings one must include the following statement: "When the diaphragm contracts it pushes out the viscera, but, in order to do so, the sternum and the ribs must resist. This resistance cannot be perfect since they are mobile and that is why, each time the diaphragm contracts, it must lift the thorax more or less. In general, the amplitude of the movement will be directly related to the resistance of the abdominal viscera and to the mobility of the ribs".

Magendie did not prove this hypothesis. He also proposed two interesting ideas:

1 - In his 1816 textbook he stated that it was clear that "the longer this muscle fiber, the greater the force that it develops".
2 - In the 1833 version of the same texbook he wrote that during normal inspiration the diaphragm was the main agonist. In a breathless patient other muscles had to be recruited, primarily the scalenes. He therefore called respiratory pulse the contraction phase of the scalenes during inspiration and proposed to use it as an index of dyspnea. This term was introduced to U.S. literature by people such as Austin Flint Jr.

After Galen, the first real investigations were made by Beau and Maissiat (1843). Using a scalpel, they cut off the scalenes, serratus and pectoralis in a dog and separated the 6th rib from the 7th from the vertebrae up to the sternum. In so doing they isolated the lower ribs from the upper ribs. They observed that the diaphragm was still inflating the lower rib cage. They believed that the pericardium was providing the fulcrum for this action.

Duchenne de Boulogne introduced electrical stimulation and used it to test previous investigators' hypotheses. His experiments, published in 1853, are now considered classic and they have been repeated and confirmed recently by P.T. Macklem and André de Troyer. In various animals and in man, Duchenne found that the diaphragm inflated the lower rib cage and that this was still true when the abdomen was open, but that it deflated the lower ribs when the animals or the dead patients were eviscerated. Therefore he stated that the diaphragm was inflating the lungs through several mechanisms: lowering the dome, lifting the ribs and pushing the lower ribs outwards.

Another fellow of Claude Bernard, Paul Bert (1870), used various devices invented by J.E. Marey or by himself. He routinely measured abdominal pressure and thoracic movement with various thoracometers.

He showed that during normal breathing abdominal pressure was positive and tracheal pressure negative.

He showed that when the intercostal accessory muscles were removed the diaphragm was inflating the lower ribs and deflating the upper ribs and that, when the animal was eviscerated, the diaphragm was deflating the lower ribs.

The common knowledge of the physiologists at the end of the 19th century is summarized in the excellent "Dictionnaire encyclopédique des sciences médicales" by Hénoque and Eloy in 1884:
- the diaphragm is the main agonist of respiration;
- it pushes down the abdominal viscera and increases the vertical diameter of the chest;
- it lifts the rib cage with its costal insertions which, because of the anatomy of the costovertebral joint and because of the direction of the ribs, pushes them upwards and outwards.

It is likely that:
- it generates a positive abdominal pressure which pushes the rib cage outwards;

- it may have an expiratory action.

Moreover, they provided additional information concerning the differences between the costal and the crural diaphragm. As I said in my introduction, we have very few concepts to add.

I do not have time here to discuss the reasons for the general oblivion of all this knowledge in the first half of the 20th century.

In fact, because respiratory muscle mechanics embraces anatomy as well as physiology, it disappeared from textbooks of physiology, not from those of anatomy. Nevertheless, it is only after the works of Moran Campbell and later of Jere Mead, J. Milic Emili, P.T. Macklem and others that the physiologists became aware of this topic once more.

However, this rediscovery is not complete and there are still several concepts which have yet to be "rediscovered".

A. GRASSINO - C. FRACCHIA - C. RAMPULLA - L. ZOCCHI
Respiratory Muscles in COPD
Bi & Gi Publishers, Verona - Springer-Verlag, London, 1988

Inspiratory muscle physiology, with particular regard to the diaphragm

MACKLEM P.T.
MEAKINS-Christie Laboratories, McGill University, Montreal, Quebec, Canada

SUMMARY: Under normal circumstances at FRC, the diaphragmatic component of respiration can be considered to be sum of the separate, parallel muscular activities of the two parts of the diaphragm. The co-ordination of these separate activities can be affected by conditions such as experimental liquid substitution of the abdominal viscera (approximating to clinical manifestations of ascites) or hyperinflation. While the respiratory role of the diagram does not suffer serious impairment in the former case, hyperinflation proves extremely damaging to inspiratory muscle function.

KEY WORDS: Inspiratory muscle physiology; diaphragm; respiration; muscular activity separated; muscular activity, parallel; abdominal viscera; ascites; hyperinflation.

As your heard from Derenne, the diaphragm is composed of two muscles: the "hamp" and the "onglet". As Derenne pointed out, Colbert stated in the 19th century that the "hamp" is essentially supplied by the upper phrenic nerve roots, the "onglet" by the lower. In English "hamp" is the costal part, and "onglet" the crural part of the diaphragm. Indeed, in dogs, if you stimulate the C5 nerve root you only record electrical activity in the costal fibers; C6 stimulation results in electrical activity in both the posterior costal and the crural regions; C7 stimulation essentially results in electrical activity in the crural part only. These experiments were performed by De Troyer[2]. This paper will describe additional experiments that Decramer et al.,[1] De Troyer et al.,[2] and Zocchi et al.,[5] have performed.

De Troyer[2] showed that the costal and the crural parts of the diaphragm have different actions when stimulated separately. Figure 1 shows motion of the rib cage (RC) plotted versus motion of the abdomen (AB) during separate stimulation of the costal and crural fibers. The dashed line is the relaxation curve obtained during passive inflation: RC expands considerably more than AB. During spontaneous quiet breathing the motions of RC and AB are essentially along the relaxation line. Stimulating the costal fibers (COS) alone, via hook electrodes inserted directly in them, results in inflation of RC and an outward displacement of AB close to, but to the right of, the relaxation line, whereas if you stimulate the crural fibers (CRU) alone, the only motion that you obtain is abdominal. CRU contraction has, in fact, no action on RC.

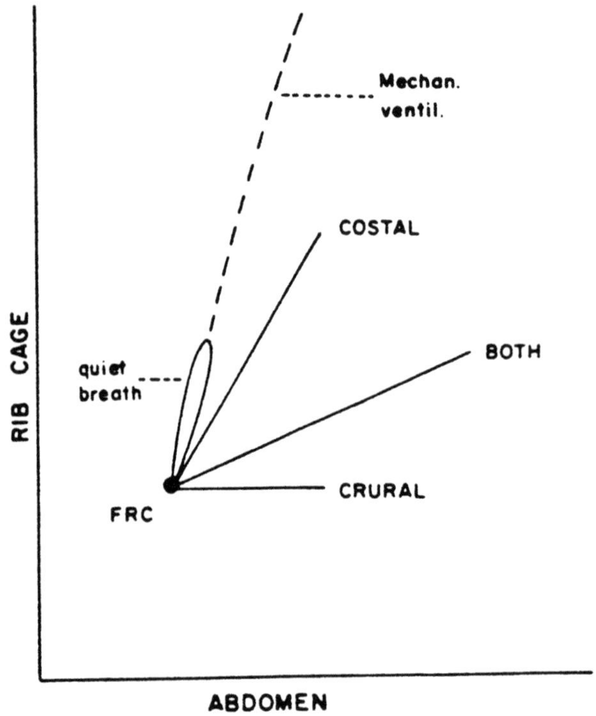

Fig. 1. - Effects of separate stimulations of costal and crural parts of the diaphragm on thoracoabdominal configuration in supine dog. Relaxation curve of the chest wall *(dashed line)* was obtained during mechanical ventilation while the animal was deeply anesthetized. [2]

When both COS and CRU are stimulated together, the net result is the sum of the vectors of COS and CRU actions; the action of the diaphragm when both COS and CRU fibers are contracting is indeed, as the earlier investigators described, inflationary on both RC and AB, but considerably displaced from the relaxation characteristics of RC and AB. The notion that the two parts of the diaphragm have different actions, different segmental innervation, different embryological and anatomical origins[2], and the fact that COS fibers are attached to RC while CRU fibers are not, have resulted in the concept that these are in fact two separate muscles. If this is accepted, it becomes very important to know how they are mechanically linked. Figure 2 is a simple mechanical model representing their mechanical linkage[3]. The respiratory system is shown in the lateral projection. RC is represented by the inverted L-shaped structure; the diaphragm is made up of COS fibers, originating from the costal margin and inserting on the rigid bar representing the central tendon, CRU fibers, originating from the vertebral column, and the medial and lateral arcuate ligaments, also inserting on the central tendon. As you can see, the two parts of the diaphragm are linked mechanically in parallel, so that when they develop force on the spring representing the lung, and elongate

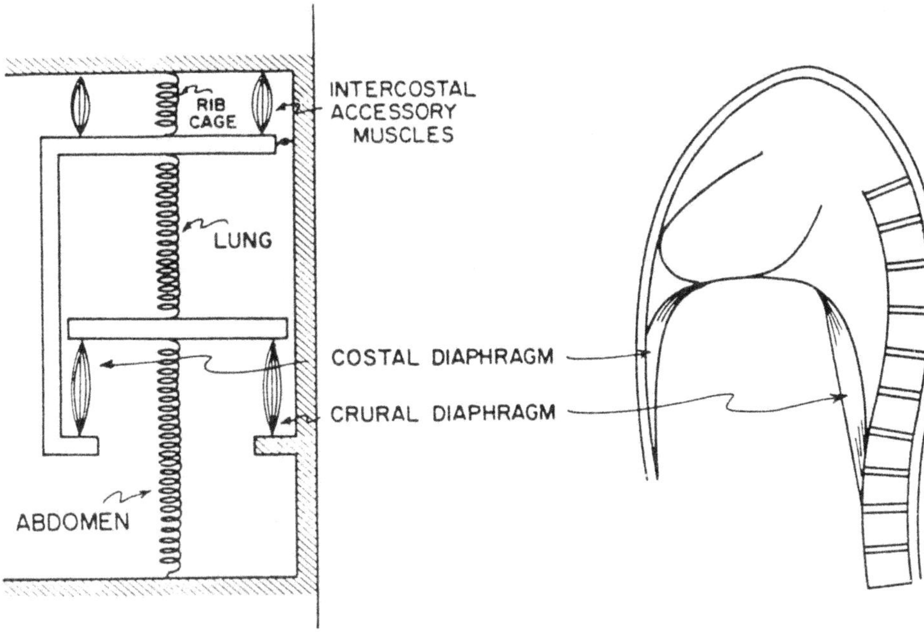

Fig. 2. - *Left*: mechanical model of inspiratory musculature. Intercostal and accessory muscles are mechanically in parallel with costal diaphragm. *Bar* into which crural and costal fibers insert represents central tendon. *L-shaped structure* represents rib cage and *spring* attached to its upper surface, elastic properties of rib cage. *Hatched area* represents rest of bony skeleton. *Right*: more anatomically realistic drawing of diaphragm illustrating separation of costal and crural parts. [3]

it, the total force they develop is the sum of the forces developed by each. Therefore, muscles arranged mechanically in parallel develop forces which are additive, and are well arranged to handle large loads. According to the same model, if COS fibers were to contract alone, they would exert an upward force on RC and an equal and opposite downward force on the central tendon: since their contraction both lifts RC and pushes abdominal content downward, it inflates the lung, pushes the AB wall out and inflates RC. On the other hand, if CRU fibers, with no connection with RC, were to contract alone, they would push AB out and elongate the spring representing the lung, but they would have no direct action on RC.

Figure 3 shows an electrical or pneumatic analogue of the mechanical model, where the different inspiratory muscles (COS, CRU and rib cage muscles) are represented as electrical generators or pumps, and the structures they displace (lung, RC, AB) are represented by rectangles. If you picture COS as a pump, which develops a negative pressure on the side towards the lung and a positive pressure on the other side, the negative pressure would inflate the lung, and the positive pressure would inflate RC and AB. For CRU, however, the negative pressure developed on one side would be transmitted through the passive COS pump and inflate the lung, while

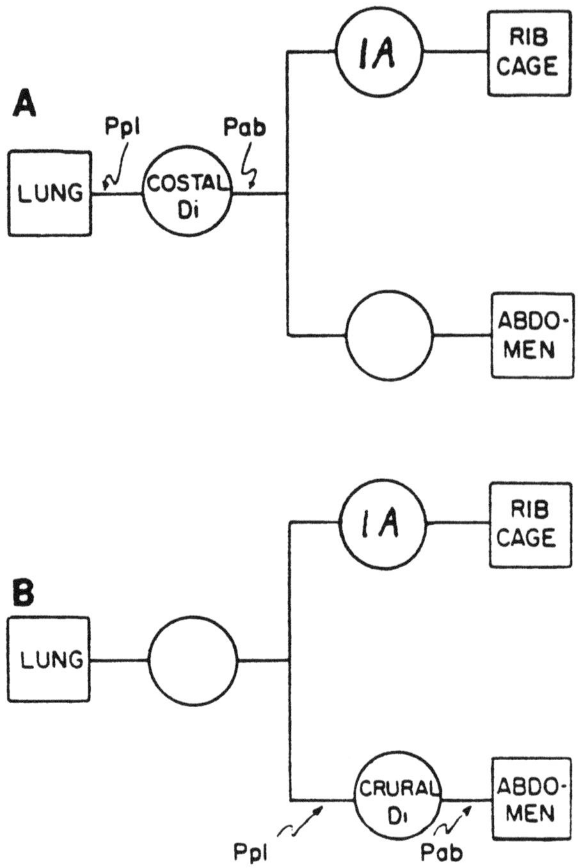

Fig. 3. - **Electrical or pneumatic analogue of mechanical model presented in Figure 2. *Top*: costal diaphragm, which, when it contracts by itself, develops a pressure on the rib cage that equals abdominal pressure. *Bottom*: crural diaphragm, which, when it contracts by itself, develops a pressure on the rib cage that equals pleural pressure. As there is no voltage (pressure) difference across generators that are inactive, abdominal pressure (Pab) is displayed on the lung side of the crural diaphragm at *top*, while pleural pressure (Ppl) is displayed on the abdominal side of the costal diaphragm at the *bottom*. I A: intercostal and accessory muscles.** [3]

the positive pressure on the other side would displace AB outward; but the negative pressure, if CRU is the only active pump (the only muscle contracting), will also be transmitted through the rib cage muscles and, according to this model, deflate RC, because a negative pressure will be applied to it. The evidence from the experimental results showed that RC did not deflate with CRU contraction. This is explained in Figure 4, where panel A shows what you have already seen in Figure 1: stimulation of CRU results in an increase in lung volume, in abdominal pressure (Pab) and in an outward displacement of AB, but there is no effect on RC. If, however, you open the abdomen to prevent any increase in Pab (panel B), RC does

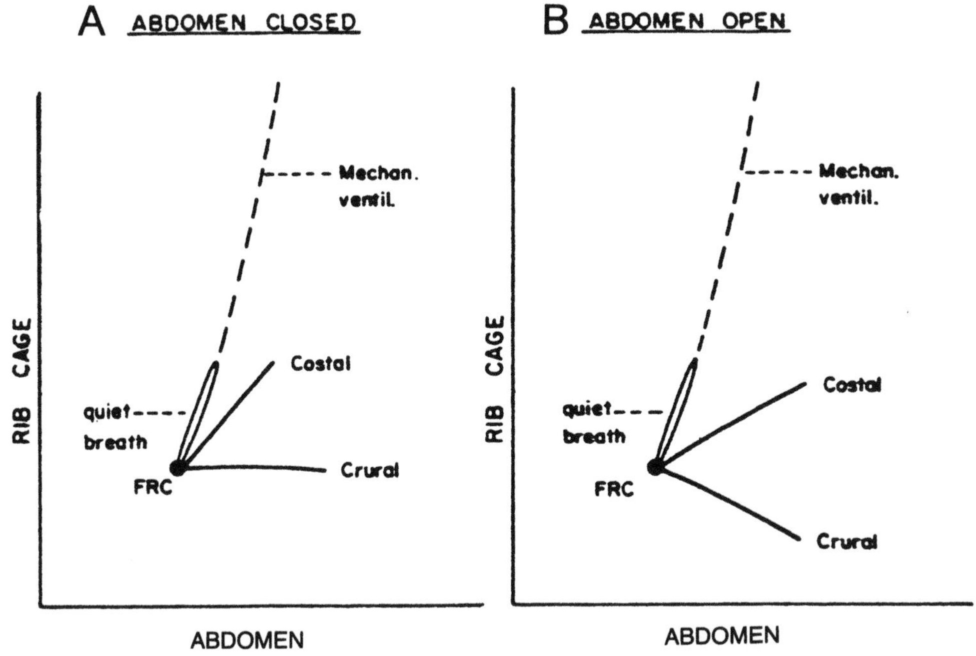

Fig. 4. - Effect of increase in abdominal pressure on changes in lower rib cage dimensions during separate stimulation of costal and crural parts of diaphragm. In panel A abdomen is closed, and abdominal pressure increases during diaphragmatic stimulation. In panel B abdomen is open, so abdominal pressure does not increase during stimulation. [2]

show paradoxical movement with CRU contraction, as predicted by the model in Figure 3. With CRU contraction and no change in Pab, the fall in pleural pressure acts to produce RC deflation, which under normal circumstances is prevented by the increase in Pab. When there is a wide pneumothorax, preventing any changes in either pleural pressure (Ppl) or Pab, the situation is restored to what it was under control circumstances. This shows that, with CRU contraction, the increase in Pab and the fall in Ppl have equal and opposite actions on the lower part of RC, preventing its movement.

Therefore, to improve our model of the action of respiratory muscles on RC and AB, we have to take into account the inflationary action of Pab and the deflationary action of the fall in Ppl on RC. This can be done as shown in Figure 5: the model is identical to that in Figure 3, with the exception that a summing junction[3] has been added, in which Ppl, Pab and the actions of COS and rib cage muscles are totalled. The net result is what determines RC motion.

The next question is: how can the two different parts of the diaphragm have different actions on RC? First of all, it must mean that when COS or CRU fibers contract, no transmission of tension occurs through the central tendon, because, if ten-

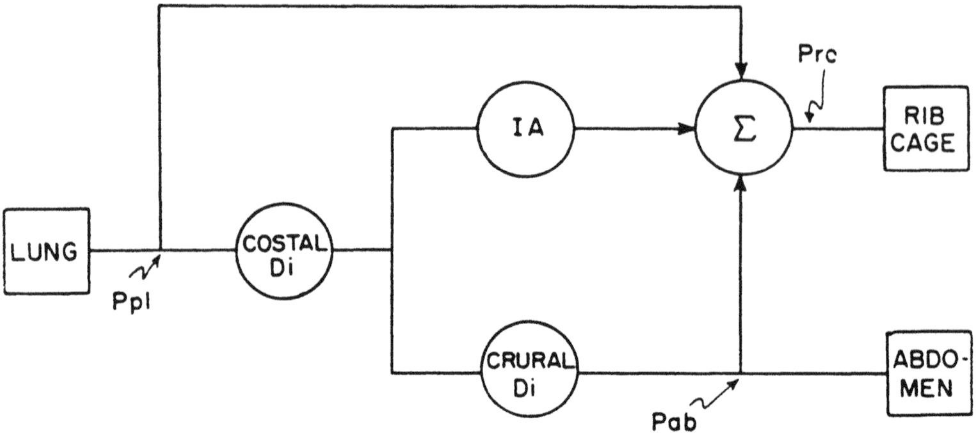

Fig. 5 - Electrical or pneumatic model of inspiratory musculature as shown in Figure 3 with addition of summing junction (Σ).[3]

sion were transmitted to COS during CRU contraction, COS would be passively stretched and, since it is attached to RC, RC would be inflated. The difference is not due to the attachment of the central tendon to the pericardium: separating the diaphragm from the pericardium has no influence on the separate actions of COS and CRU. Furthermore, if with CRU contraction there were everywhere in the abdomen an equal increase in Pab, and everywhere over the pleural surface an equal fall in Ppl, there would be a pressure difference across COS, displacing it upwards into the thorax, again passively stretching it and thereby lifting RC. This led Decramer[1] to think that, in order to account for the different actions of COS and CRU, there had to be regional differences in Pab. Indeed, Figure 6 shows a plot of Pab under CRU versus Pab under COS. The line of identity is shown, the closed circles representing Pab with CRU stimulation: all data points fall below the line of identity, indicating that Pab during CRU stimulation is considerably greater under CRU than it is under COS. When COS fibers are stimulated alone, as shown by the closed squares, all data points fall above the line of identity, indicating that the Pab swings under COS are considerably greater than under CRU. Therefore, with CRU contraction Pab is not perfectly transmitted under the COS fibers, accounting for the fact that CRU contraction does not have an inflationary action on RC by passively stretching the COS fibers. The failure of transmission of Pab permits the two muscles to act as if they were mechanically in parallel. Subsequently Zocchi[5] showed that when a large volume of abdominal viscera (the spleen and a large part of the small intestine) is removed and the abdomen filled with an equal volume of liquid (liquid substitution) and subsequently closed, the Pab swings (open symbols) fall along the line of identity, irrespective of which part of the diaphragm

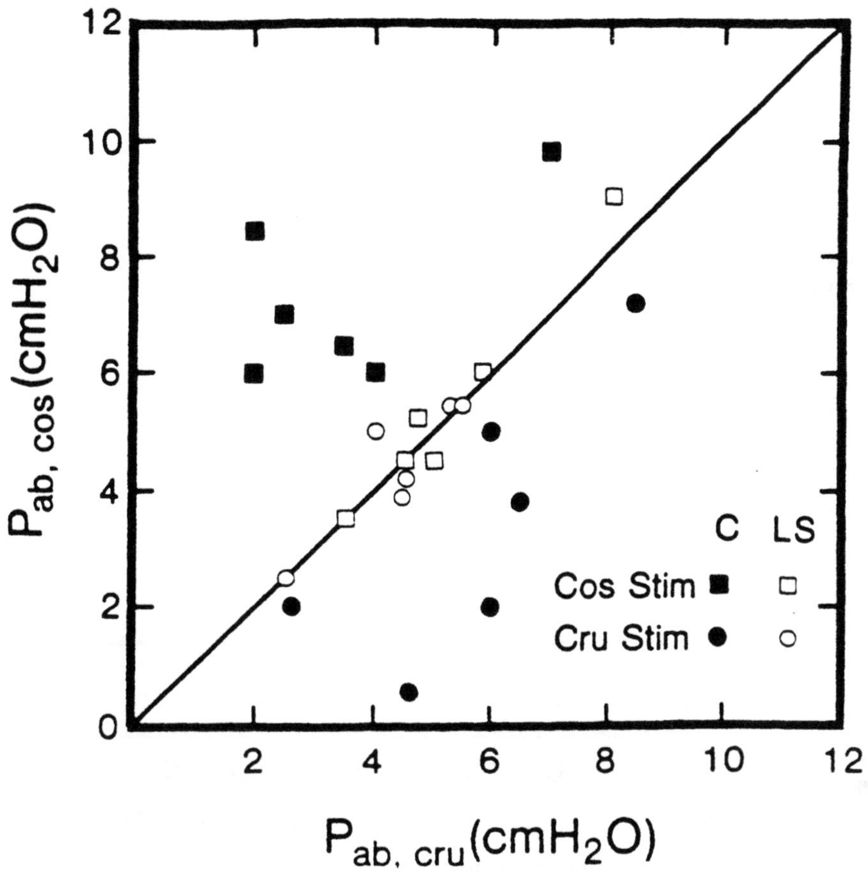

Fig. 6 - Comparison of pressure swings recorded under costal (Pab, cos) a crural (Pab, cru) diaphragms during costal (*squares*) and crural (*circles*) stimulations. *Filled symbols* are measurements obtained under control conditions, *open circles* measurements obtained when stimulations were repeated in the same 6 dogs after liquid-substitution of part of abdominal contents. *Solid line* is identity line. One measurement in each condition is shown for each animal.[5]

is contracting. Under these circumstances the two parts of the diaphragm have identical actions on RC. This is shown in Figure 7. On the ordinate is the ratio of RC to AB motion: positive values mean that both RC and AB are displaced in the inflationary direction; a value of zero means that there is no movement of RC, while AB moves outwards; negative values mean that RC paradoxes, while AB moves outwards.

Under control circumstances, with COS stimulation the value is positive (i.e. both RC and AB are displaced in the inflationary direction), whereas with CRU stimulation there is no motion of RC, and only AB is displaced. However, with liquid substitution and equalization of Pab, the actions of both parts of the diaphragm become identical, both matching the action of COS under control circumstances.

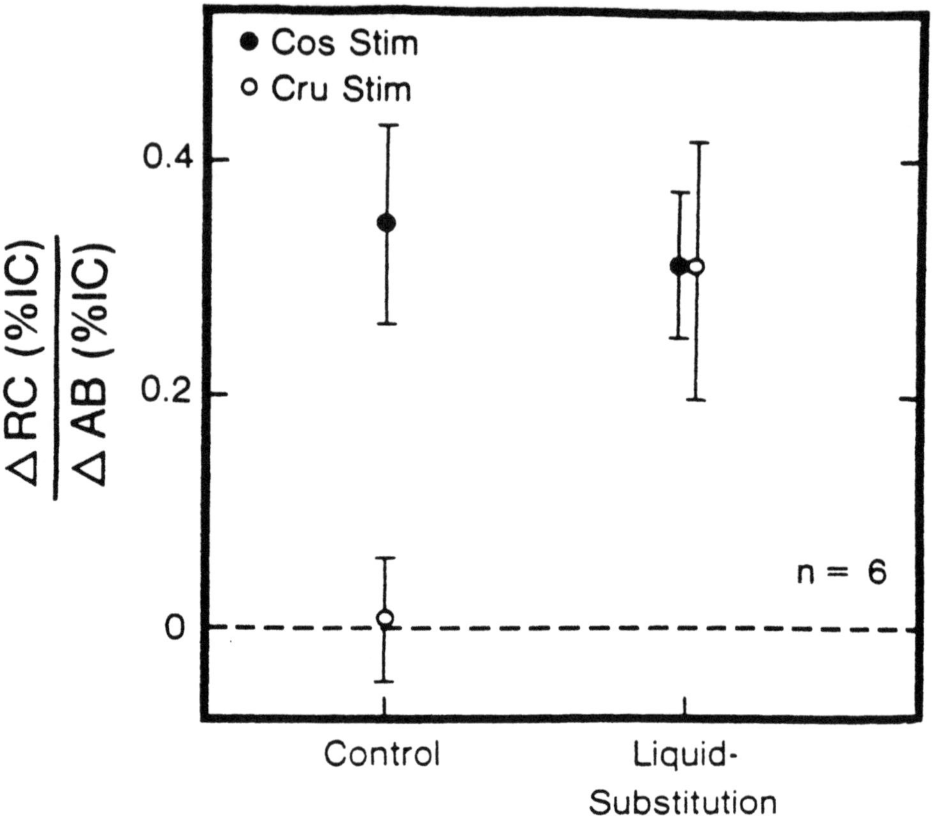

Fig. 7 - Effect of liquid-substitition on action of COS (*filled symbols*) and CRU (*open symbols*) stimulations on chest wall configuration. Values are mean (± SE) slopes on configurational changes from FRC observed in 6 dogs.[5]

CRU has now been converted to a muscle which inflates RC by increasing Pab under the costal part of the diaphragm, so that the model changes from two muscles arranged in parallel to two muscles connected in series by a pulley (as first suggested by Peterson and Otis)[4]. In this arrangement, the tension developed by one muscle is transmitted to the other via the pulley, but the force applied by the muscles to the structures they displace is the sum of the forces developed by each muscle.

Let us now consider hyperinflation, because this has a major impact on the action of the diaphragm and on the mechanical linkage between COS and CRU. Because this symposium deals with respiratory muscles in COPD, hyperinflation is quite important. Figure 8 shows results obtained in 6 dogs by Zocchi[5], for RC versus AB motion with COS and CRU stimulation at FRC and at various lung volumes above FRC. As already shown, at FRC, COS stimulation inflates both RC and AB, while CRU stimulation only displaces AB; with hyperinflation of relatively mild degree, however, the inflationary action of COS fibers on RC disappears,

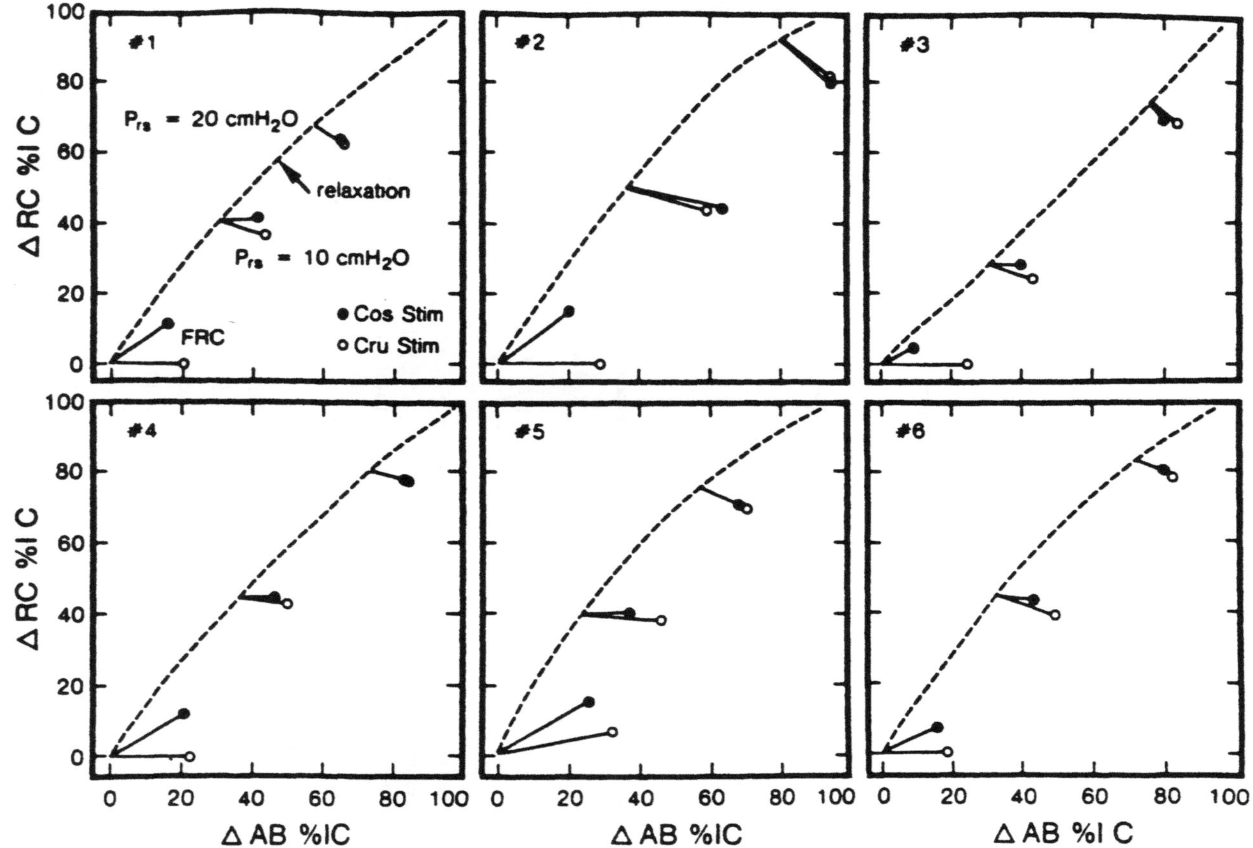

Fig. 8 - Individual plots of relative RC-AB motions in 6 supine dogs. Displacements are plotted as percent of inspiratory capacity (IC). *Broken lines* are relaxed relationships obtained during passive inflations; *solid lines* show RC-AB displacements resulting from COS (*filled circles*) and CRU (*open circles*) stimulations at chest wall configurations corresponding, in each dog, to FRC, Prs = 10 cmH_2O and Prs = 20 cmH_2O. Each point is mean of 3-5 measurements. Differences between slopes resulting from COS and CRU stimulations were reduced at the intermediate level of hyperinflation and abolished at the highest level, associated with progressively larger departures from relaxation lines.[5]

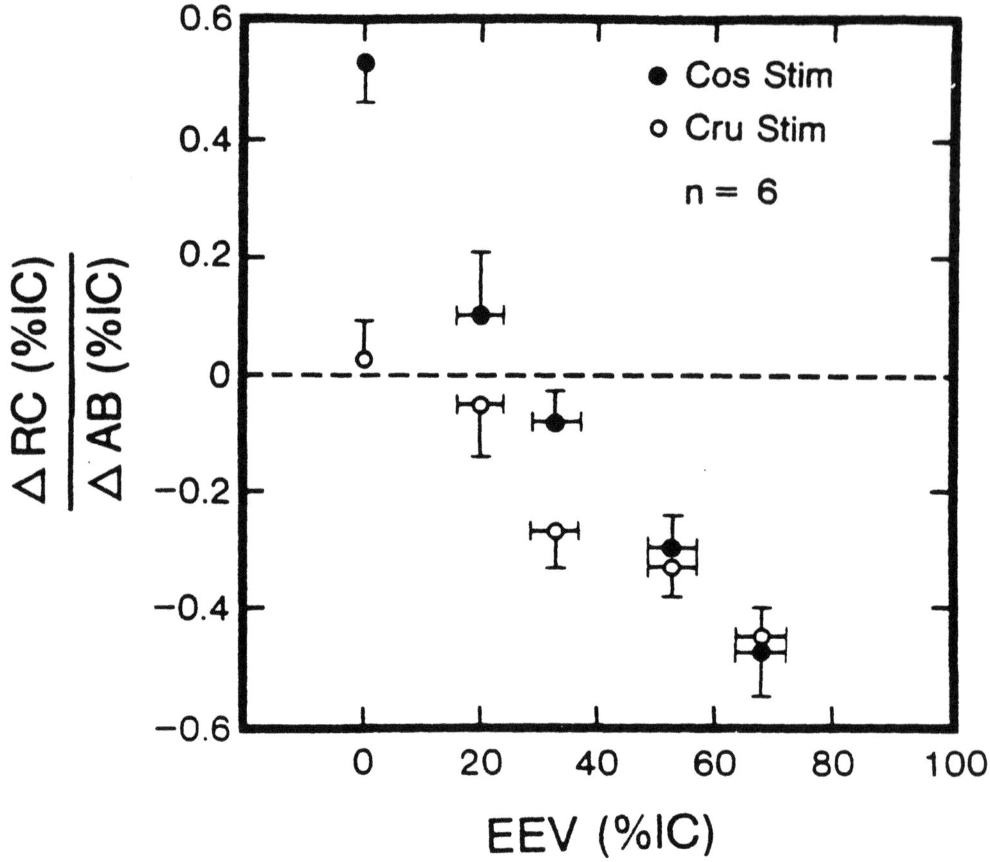

Fig. 9 - Influence of lung volume on action of COS (*filled circles*) and CRU (*open circles*) stimulations on chest wall configuration. Mean values (± SE) of slopes of configurational changes are plotted as functions of mean (± SE) end-expiratory volume (EEV % IC) obtained at CPAP of 0, 5, 10, 15, and 20 cmH_2O. Data from 6 dogs. Data points above or below the *dashed line* (slope of O = no RC displacement) indicate, respectively, RC inflation or deflation.

and in fact the two parts of the diaphragm have virtually identical actions. At higher lung volumes both parts become deflationary on RC. This is summarized in Figure 9, showing the ratio of RC to AB motion as a function of inspiratory capacity from FRC to TLC: again, a positive value indicates that both RC and AB are being inflated; a value of zero indicates no motion of RC; a negative value means paradoxical motion of RC, as AB is being displaced outwards. Only small increases in lung volume, within the tidal volume range, decrease the inflationary action of COS fibers on RC dramatically, and CRU fibers begin to become deflationary: under these circumstances it becomes difficult to breathe. Now both muscles of the diaphragm have the same action, just as they did with liquid substitution, but instead of both them being inflationary to RC, both are now deflationary on it. Figure 10A is the

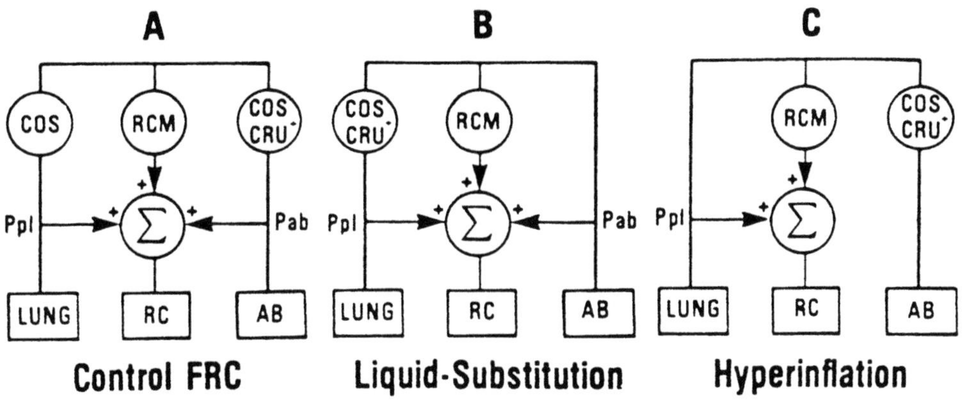

Fig. 10 -Electric or pneumatic model of inspiratory muscles. A) FRC: same model shown in Figure 5 in addition, COS is partly in parallel and partly in series with CRU. Pab (via the area of apposition) and Ppl act on RC depending on their magnitudes and gains on the summing junction (Σ). The plus signs at the summing junction indicate that the junction's function is to compute the algebraic sum of its inputs.
B) Liquid-substitution of part of abdominal contents: COS and CRU are entirely in parallel, and they can be represented by a single generator or pump in series between both lungs and RC and lungs and AB, and in series with RCM.
C) Hyperinflation: COS and CRU are entirely in parallel and can be represented by a single generator or pump in series between lung and AB, but in parallel between lung and RC, and in parallel with RCM. Area of apposition is abolished, so that Pab generated by diaphragmatic contraction no longer acts on RC.[5]

model you have already seen: with liquid substitution (Fig. 10 B), CRU fibers move into a situation where they actively inflate both RC and AB, and the diaphragm can now be portrayed as a single muscle, developing a positive pressure on one side, which inflates both RC and AB, and a negative pressure on the other side, which inflates the lung. But with hyperinflation (Fig. 10 C), both parts of the diaphragm have a deflationary action on RC, so that the appropriate representation is a single muscle which develops a negative pressure on one side to both inflate the lung and deflate RC, while the only inflationary action on the chest wall is a positive pressure on the abdominal side, acting to inflate AB. If this model is accurate, the only way a hyperinflated individual can prevent paradoxical RC displacement with each inspiration is to recruit his RC muscles, in order to develop a positive pressure on RC. This, as can intuitively be seen, is a very inefficient way of breathing, and is one of the major reasons why hyperinflation is so damaging to inspiratory muscle function.

To summarize, the diaphragm under normal circumstances at FRC can be considered as two muscles acting in parallel, with separate actions. The linkage between these two muscles changes in different ways in different circumstances: with liquid substitution, for example (and this can clinically be represented by ascites), and with hyperinflation. In the former instance, inspiratory muscle function is not affected to any serious degree, but with hyperinflation the action of the diaphragm is seriously impaired.

References

1. DECRAMER M., DE TROYER A., KELLY S., ZOCCHI L., MACKLEM P.T.: Regional differences in abdominal pressure swings in dogs. J. Applied Physiol., 1984, 57, 1682-87.
2. DE TROYER A., SAMPSON M., SIGRIST S., MACKLEM P.T.: Action of costal and crural parts of the diaphragm in dogs. J. Applied Physiol., 1982, 53, 30-39.
3. MACKLEM P.T., MACKLEM D.M., DE TROYER A.: A model of inspiratory muscle mechanics. J. Applied Physiol., 1983, 55, 547-557.
4. OTIS A.B., PETERSON, C.V., J.R.: Some propositions concerning mechanics of the diaphragm. Fed. Proceed., 1984, Abstract. 43,529.
5. ZOCCHI L., GARZANITI N., NEWMAN S., MACKLEM P.T.: Effect of hyperinflation and equalization of abdominal pressure swings on diaphragm action. J. Applied Physiol., 1987, 62, 1655-1664.

A. GRASSINO - C. FRACCHIA - C. RAMPULLA - L. ZOCCHI
Respiratory Muscles in COPD
Bi & Gi Publishers, Verona - Springer-Verlag, London, 1988

Respiratory muscle physiology with particular regard to rib cage muscles

M. ESTENNE
Chest Service, Erasme University Hodpital, Brussels, Belgium

SUMMARY: The mechanism responsible for the highly effective coupling of separate rib cage and abdomen volume changes during breathing is of great interest to respiratory physiologists.
Some investigators have postulated the diaphragm as the sole muscular agent recruited in quiet inspiration. A review of recent studies in dogs and humans challenges this view.
The author, on the basis of personal observation of trained human subjects, concludes that diaphragm activity is co-ordinated with that of the parasternal intercostals and scalenes to confer considerable displacement to the chest wall during quiet breathing.

KEY WORDS: Respiratory muscles physiology; rib cage muscles; abdomen volume changes; breathing; respiratory physiologists; diaphragm; parasternal intercostals; scalenes; chest wall.

Since the studies of Konno and Mead[1] on the separate volume changes during breathing of the rib cage and abdomen, it has been recognized that quiet breathing in upright humans generally occurs on the relaxed thoracoabdominal configuration. Under these circumstances, not only does the chest wall (i.e., the rib cage - abdomen system) move on its relaxation characteristic, but the rib cage also behaves with a singular degree of freedom: during inspiration, all rib cage diameters increase in the same proportion relative to the relaxed rib cage configuration. The mechanism of this extremely effective "coupling" between the rib cage and abdomen has been a matter of considerable interest among respiratory physiologists, in particular since Goldman and Mead[2] published their famous hypothesis. These authors postulated that in upright humans breathing quietly the diaphragm is the only important contracting muscle, driving the entire chest wall, including the rib cage, on its relaxed configuration. In other words, they suggested that there is no significant agonistic activity of the rib cage muscles during quiet inspiration.

This hypothesis, however, has been challenged by studies performed in patients with complete transection of the lower cervical cord, which leaves the diaphragm intact but causes paralysis of all the intercostal and abdominal muscles, and occasionally of the scalene muscles[3, 4]. In such patients, the tidal volume loop departs substantially from the chest wall and rib cage relaxation characteristic: the rib cage expansion is invariably reduced relative to the abdominal expansion, the lower rib cage expands proportionately more along its transverse than anteroposterior diameter, and the upper rib cage moves paradoxically inward during inspiration. In addition, studies in patients with transection of the upper cervical cord have shown that the chest wall is markedly distorted during isolated contraction of the paced diaphragm[5]. All these observations are incompatible with Goldman and Mead's hypothesis[2], and suggest that the diaphragm acting alone is not able to drive the chest wall on its relaxed configuration. To do so, it needs a good deal of help from other inspiratory muscles which contribute to expand the rib cage during quiet breathing. Recent work in this area suggests that most of this help is provided by the parasternal intercostals and the scalene muscles[6, 7].

Since Taylor's electrical recordings from the intercostal muscles in humans[8], it is well established that the interchondral portion of the internal intercostals (the "parasternals") is electrically active during quiet inspiration[9, 10, 11], and recent studies have shown that this EMG activity cannot be voluntarily suppressed unless tidal volume becomes negligible (less than 100 ml)[10]. In line with Goldman and Mead's hypothesis[2], the parasternal EMG activity has been initially interpreted as being the consequence, rather than the cause, of the rib cage displacement[12]; it is seen as probably reflexive in nature, occurring through activation of the spindle mechanism to prevent the rib cage distortion that would otherwise result from the fall in pleural pressure produced by diaphragmatic contraction. Recent studies, however, have shown that the parasternals are poorly supplied with muscle spindles[13]. In addition, the parasternal EMG activity is reduced or even abolished during breathing against resistive or elastic inspiratory loads, even though the tidal fall in pleural pressure is increased under these circumstances[14]. Similarly, this activity is also reduced in the presence of rib cage distortions produced by diaphragmatic isovolume maneuvers[10]. All these observations suggest that the EMG activity detected in the parasternals is not reflexive in nature and does not relate to the fall in pleural pressure. This conclusion has been recently strengthened by studies in dogs which have shown that the contraction of the parasternals is a real myometric contraction[15]; these muscles actively shorten during inspiration and lengthen during expiration. Although the length changes undergone by the parasternals during breathing in humans have not yet been measured, it thus appears that the parasternals are real agonists (as opposed to fixators) and actively contribute to inspiration.

Phrenicotomized dogs provide a good experimental setting for assessing the mechanical action of the parasternals on the rib cage because, after bilateral section of the phrenic nerve roots in the neck, the parasternals are thought to be the only important inspiratory muscles still in activity[16]. Studies in anesthetized dogs after bilateral phrenicotomy have demonstrated that contraction of the parasternals causes the rib cage to expand more along its transverse than its anteroposterior diameter relative to its relaxed configuration. In addition, the rib cage is markedly distorted along its axial (cephalocaudal) axis. Indeed, in the phrenicotomized dog, the ribs move in a cranial direction, whereas the sternum is displaced in a caudal direction. This is due to the insertions of the parasternals on the lateral border of the sternum and the upper border of the costal cartilages, which produce a rotation of the chondrosternal junction when they contract[16].

To investigate the mechanical action of the parasternals on the human rib cage, we have studied the pattern of rib cage motion in four subjects who were highly trained in respiratory maneuvers and were able to virtually abolish diaphragmatic use during tidal breathing[17]. This maneuver elicited a marked increase in the parasternal EMG activity, while there was no significant change in the scalene inspiratory EMG activity. In addition, the sternocleidomastoids, the pectoralis major, and the abdominal muscles were always electrically silent. We observed that this pattern of respiratory muscle use produced rib cage deformations that were very similar to those seen in dogs after phrenicotomy. The rib cage expanded considerably more along its transverse than anteroposterior dimension relative to its relaxed configuration[17]. It thus appears that when the parasternals contribute most of the tidal volume, they expand predominantly the lateral walls of the cage ("bucket-handle" action) and they displace the sternum in the caudal direction. During quiet inspiration in normal humans, however, the sternum moves in a cranial, not a caudal, direction. This suggests that in man there is an additional muscle or muscle group which is active during quiet breathing and opposes the action of the parasternals on the sternum. This muscle group is made up of the scalene muscles.

The scalenes, together with the sternocleidomastoids, have traditionally been described under the heading "accessory muscles", because it was thought that they were inactive at rest, being recruited only when ventilation increases. However, the initial observations by Campbell[18] in normal humans were obtained with skin surface electrodes, a technique which is generally unable to recognize low levels of EMG activity. And indeed, using needle electrodes directly inserted into the muscle, a number of investigators have subsequently documented a phasic inspiratory EMG activity in the scalenes of normal subjects breathing at rest[9,19,20]. This activity has a time course which closely resembles that of the diaphragm and parasternals, and cannot be voluntarily suppressed as long as tidal volume is kept within reasonable limits[11]. There is thus no reason for using the qualifying adjective "accessory" in describing the scalenes. These muscles in humans are primary, not accessory, muscles of inspiration[11].

The mechanical action of the scalenes in humans is poorly defined, presumably because there is no clinical setting that spares the scalenes and causes paralysis of all the other inspiratory muscles. In the dog, however, selective stimulation of the scalenes is primarily associated with an axial displacement of the ribs and the sternum in a cranial direction[21]. This cranial displacement of the ribs, in turn, causes the rib cage to expand along both the anteroposterior and transverse diameters, but the rib cage expansion is proportionately greater along its anteroposterior than transverse diameter ("pump-handle" action). In man, recent evidence has been obtained which suggests that the scalenes are actively involved in expanding the upper rib cage along its anteroposterior diameter during quiet breathing. Depending on the level of the cervical cord transection, tetraplegic patients may show a complete paralysis of the scalenes or, alternatively, may retain a phasic inspiratory EMG activity in these muscles[4]. We have observed that, in such patients, the absence of scalene phasic EMG activity is almost invariably associated with an inward inspiratory displacement of the upper rib cage, whereas in general there is no paradox when the scalenes are phasically active during inspiration. Similarly, recent studies have shown that the upper rib cage moves inward when normal subjects attempt to inspire with the diaphragm alone and, in so doing, selectively inhibit the scalenes[11]. It thus appears that the EMG activity present during inspiration in the scalenes is not simply an electrical event; the scalenes normally contribute to quiet inspiration in normal humans, acting to counteract the parasternals in their action on the sternum and to expand the upper rib cage primarily along its anteroposterior dimension.

In conclusion, the studies reported here indicate that: 1) the diaphragm acting alone is not able to drive the entire chest wall along its relaxed configuration; 2) the human rib cage is more distortable than conventionally thought, its inherent structural stability not being sufficient to ensure its behaviour as a unit; 3) when it moves with a singular degree of freedom during quiet breathing, it is because the diaphragm, the parasternal intercostals and the scalenes are activated in concert in a coordinated manner.

References

1. KONNO K., MEAD J. Measurement of the separate volume changes of rib cage and abdomen during breathing. J. Appl. Physiol. 1967, 22: 407-22.
2. GOLDMAN M.D., MEAD J. Mechanical interaction between the diaphragm and rib cage. J. Appl. Physiol. 1973, 35: 197-204.
3. MORTOLA J.P. SANT'AMBROGIO G. Motion of the rib cage and the abdomen in tetraplegic patients. Clin. Sci. Mol. Med. 1978, 54: 25-32.
4. ESTENNE M., DE TROYER A. Relationship between respiratory muscle EMG and rib cage distortion in tetraplegia. Am. Rev. Respir. Dis. 1985, 132: 53-9.
5. DANON J., DRUZ W.S., GOLDBERG N.B., SHARP J.T. Function of the isolated paced diaphragm and the cervical accessory muscles in C1 quadriplegics. Am. Rev. Respir. Dis. 1979, 119; 909-19.
6. DE TROYER A. Actions of the respiratory muscles or how the chest wall moves in upright man. Bull. Eur. Physiopathol. Respir. 1984, 20: 409-13.
7. DE TROYER A., LORING S.H. Actions of the respiratory muscles. In: Handbook of physiology, section 3: Respiration, vol. 2: Mechanics of breathing. American Physiological Society, Bethesda, 1984, pp 443-61.

8. TAYLOR A. The contribution of the intercostal muscles to effort of respiration in man. J. Physiol (London) 1960, 151: 390-402.
9. DELHEZ L. Contribution electromyographique à l'étude de la mécanique et du contrôle nerveux des mouvements respiratoires de l'homme. Thèse, Vaillant-Carmanne, Liege, 1974.
10. DE TROYER A., SAMPSON M. Activation of the parasternal intercostals during breathing efforts in human subjects. J. Appl. 4 Physiol. 1982, 52: 524-9.
11. DE TROYER A., ESTENNE M. Coordination between rib cage muscles and diaphragm during quiet breathing in humans. J. Appl. Physiol. 1984, 57: 899-906.
12. DERENNE J.P., MACKLEM P.T., ROUSSOS C. The respiratory muscles: mechanics, control and pathophysiology. Am. Rev. Respir. Dis. 1978, 118: 373-90.
13. DURON B. Intercostal and diaphragmatic muscle endings and afferents. In: Lung Biology in Health and Disease, vol. 17: Regulation of breathing, part I. Hornbein T.F. ed. Dekker, New York, 1981, pp 473-540.
14. SAMPSON M.G., DE TROYER A. Role of intercostal muscles in the rib cage distortions produced by inspiratory loads. J. Appl. Physiol. 1982, 52: 517-23.
15. DECRAMER M., DE TROYER A. Respiratory changes in parasternal intercostal length. J. Appl. Physiol, 1984, 57: 1254-60.
16. DE TROYER A., KELLY S. Chest wall mechanics in dogs with acute diaphragm paralysis. J. Appl. Physiol. 1982, 53: 373-9.
17. DE TROYER A., ESTENNE M., NINANE V. Rib cage mechanics in simulated diaphragmatic paralysis. Am. Rev. Respir. Dis. 1985, 132: 793-9.
18. CAMPBELL E.J.M. The role of the scalene and sternomastoid muscles in breathing in normal subjects. An electromyographic study. J. Anat. 1955, 89: 378-86.
19. RAPER A.J., THOMPSON W.T. Jr., SHAPIRO W., PATTERSON J.L. Jr. Scalene and sternomastoid muscle function. J. Appl. Physiol. 1966, 21: 497-502.
20. JONES D.S. BEARGIE R.J., PAULY J.E. An electromyographic study of some muscles of costal respiration in man. Anat. Record 1953, 117: 17-24.
21. DE TROYER A., KELLY S. Action of neck accessory muscles on rib cage in dogs. J. Appl. Physiol. 1984, 56: 326-32.

A. Grassino - C. Fracchia - C. Rampulla - L. Zocchi
Respiratory Muscles in COPD
Bi & Gi Publishers, Verona - Springer-Verlag, London, 1988

Patterns of inspiratory muscle activation with changing levels of ventilatory demand

E. D'Angelo
Institute of Human Physiology, University of Milan, Italy

SUMMARY: Electrical activation of inspiratory muscular activity in response to changing ventilatory demand has been examined by means of implanted electrodes in vagotomized and anaesthetized animals. Findings suggest that the response elicited ultimately depends on the role of the central inspiratory motor control system.

In postulating the applicability of such findings to human subjects, the greater recruitment of extra-diaphragmatic than diaphragmatic inspiratory muscle components observed in anaesthetized animals tallies with the greater contribution of rib cage expansion to tidal volume recorded during CO_2 rebreathing in man.

However, the paucity of relevant human data and the possible conditioning of animals responses by anaesthesia mean that acceptance of this hypothesis must as yet be subject to reservations.

KEY WORDS: Inspiratory muscle activation; ventilatory demand; vagotomized animal; anaesthetized animal; inspiratory control system; rib cage expansion; tidal volume.

Rapid inspirations in man are characterized by a greater contribution of rib cage expansion to any lung volume increment over most of the inspiratory capacity[5, 25]. Enhancements of rib cage contribution to tidal volume also occur when ventilation is increased in normal subjects by exercise or CO_2 rebreathing[16, 19]. Although the features of chest wall motion with increasing effort of inspiration could depend in part on the mechanical characteristics of the relaxed respiratory system[4, 14, 15], these observations suggest that the distribution of activation between the inspiratory muscles may differ according to the ventilatory demand. In this respect data in man are scanty and not suitable for quantitative analysis. Moreover, surface electromyography is usually performed in humans, and some technical limitations might be inherent to surface EMG recording as far as inspiratory muscles are concerned[13].

In a series of studies on rabbits, cats and dogs[6, 7, 8, 9], electrical activity has been simultaneously recorded via implanted bipolar electrodes from various inspiratory muscles and quantitated as the moving time average EMG (time constant: 42 ms) under a variety of experimental conditions, with the aim: 1) of assessing the pattern

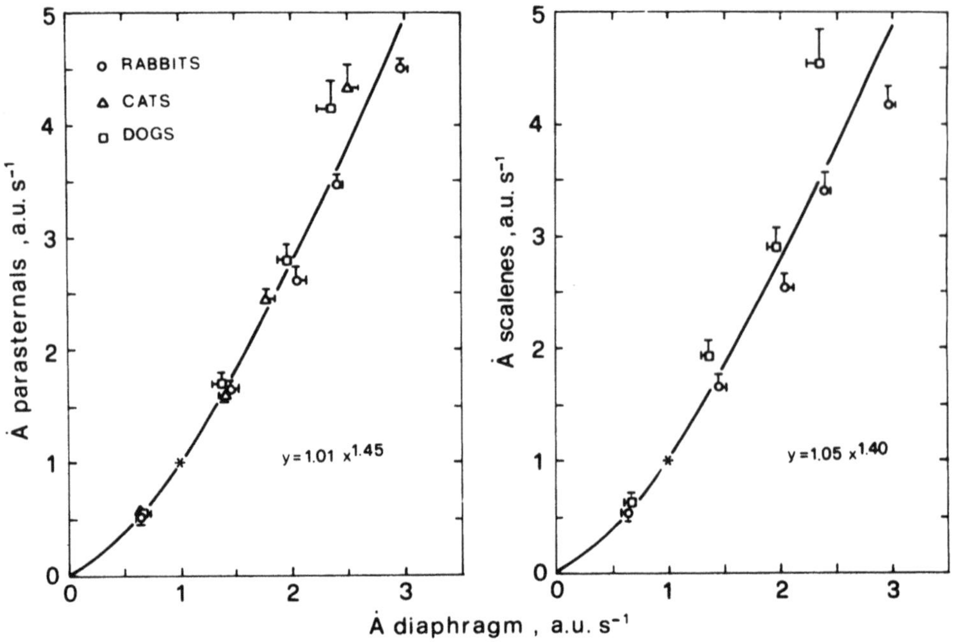

Fig. 1. - The relationship between mean rate of rise of the moving average EMG (A) of the diaphragm and that of the parasternal or the scalene muscles during rebreathing in various animal species (see key to symbols). Reference values of mean rate of rise of electrical activity for each muscle are those obtained during eupneic breathing (asterisk). Data from D'Angelo and co-workers[6, 7, 8, 9], plus unpublished observation; data points are best fitted by a power function with exponent significantly greater than unity. Bars are 1 SE.

of inspiratory muscle activation when the ventilatory demand is made to change; and 2) of elucidating the mechanism involved in the response of the various inspiratory muscles.

In lightly anesthetized rabbits, cats and dogs during eupneic breathing, the parasternals and the scalenes appear to be the only active inspiratory muscles besides the diaphragm. Moreover, activity reappears simultaneously in these muscles on resumption of breathing from apnea by artificial overventilation, increasing together with the effort of inspiration. These muscles can therefore be classified as primary inspiratory muscles, and our investigation has been focused on them.

During rebreathing, changes in both the peak and the rate of rise of the moving average EMG are systematically larger for the scalenes and the cranial and caudal parasternals than for the diaphragm (Fig. 1). Other conditions modify the ventilatory demand, besides chemical stimulation; mean inspiratory flow increases with hyperthermia (body temperature $\geqslant 40°C$) and passive motion of the hind limbs, whereas it decreases with hypothermia (body temperature $\leqslant 35°C$) and deeper levels of anesthesia. For these conditions too, the changes in extra-diaphragmatic inspiratory muscle activity are relatively larger than those in diaphragm activity. Indeed, in rabbits the mean rate of rise of the moving average EMG with hyperthermia and passive limb motion increases relatively more for the parasternal muscles than for the diaphragm, whereas with hypothermia and deeper barbiturate anesthesia the same parameter decreases relatively more in the parasternals than the diaphragm[8].

It appears therefore that 1) with increasing ventilatory demand there is a relatively greater increase in the activity of the extra-diaphragmatic inspiratory muscles than in that of the diaphragm; and 2) the relationship between the relative changes in diaphragm and in extra-diaphragmatic inspiratory muscle activity is largely independent of the condition that modifies the ventilatory demand.

Differences in the response to enhanced respiratory drive also occur between the scalenes and the cranial and caudal parasternals, but they are small as compared to those between any of these muscles and the diaphragm. In rabbits and dogs there is a tendency for the mean rate of rise of the moving average EMG to increase during rebreathing more in the scalenes and cranial parasternals than in the caudal parasternals[6, 9]. In any case, changes in both peak and rate of rise of the moving average EMG of these muscles are linearly related to each other, and the slope of these relationships is in general close to unity.

Extra-diaphragmatic inspiratory muscle activity appears to be largely independent of mechanical factors which might have caused autogenetic facilitation or inhibition via muscle spindles or tendon organs. Indeed, substantial changes in chest wall configuration and in the pattern of chest wall motion occur in animals between the supine and the head-up posture[3]; the operating length of the inspiratory muscles and the tension they develop for a given degree of activation are also changed, with shortening of the diaphragm and lengthening of the extra-diaphragmatic inspiratory muscles, in the head-up relative to the supine position[11, 12, 18]. In spite of this, the relationships between the relative changes in diaphragm activity and those of any of the extra-diaphragmatic inspiratory muscles obtained during rebreathing in dogs are essentially the same in both postures[9]. Even more marked changes in the pattern of chest wall motion occur with diaphragm paralysis; but again parasternal muscle activity remains unaltered when complete blockade of conduction in the phrenic nerves is performed during single breathing cycles in vagotomized rabbits, independently of the prevailing level of respiratory drive[6].

Vagal afferent signals substantially affect inspiratory muscle activation. While the rate of rise of the moving average EMG of the diaphragm remains unchanged with bilateral vagotomy, that of the extra-diaphragmatic inspiratory muscles is reduced by about one third. This occurs in all the species studied, irrespective of the level of chemical drive[7, 8, 9]. Hence vagal signals exert selective facilitatory effects on the extra-diaphragmatic inspiratory muscles. Moreover, the vagal-dependent facilitation is essentially the same for the scalenes and the cranial and caudal parasternals. As a consequence, the relationship between diaphragmatic and extra-diaphragmatic inspiratory muscle activity during rebreathing shifts downwards and rightwards in vagotomized animals (Fig. 2). The same occurs in intact animals with increasing

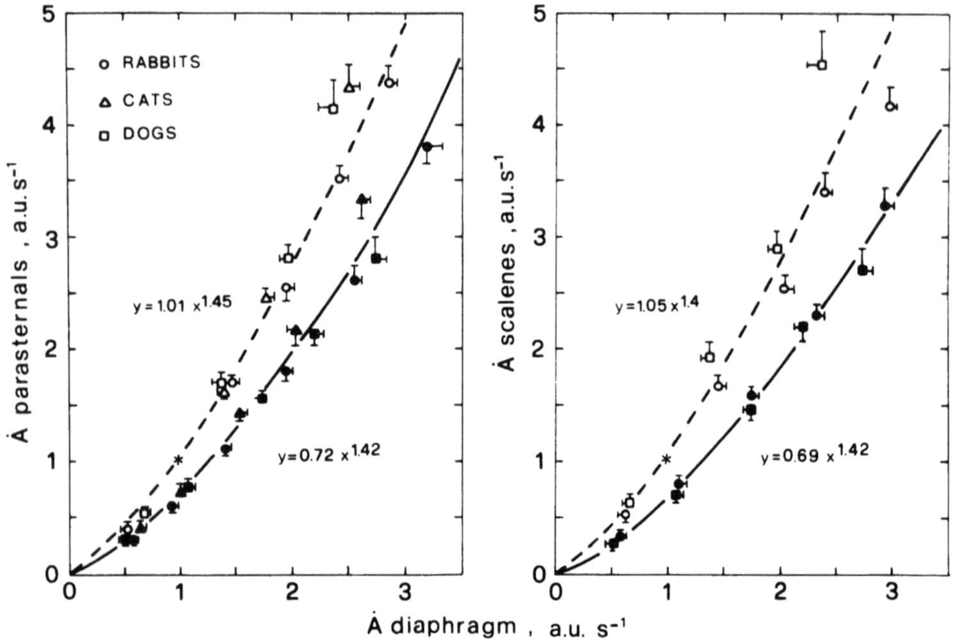

Fig. 2. - The relationship between mean rate of rise of the moving average EMG (A) of the diaphragm and that of the parasternal or the scalene muscles during rebreathing in various animal species (see key to symbols) before and after bilateral vagotomy (open and closed symbols, respectively). Reference values for each muscle are those obtained during eupneic breathing before vagotomy (asterisk). Data from D'Angelo and co-workers[6, 7, 8, 9], plus unpublished observations; bars are 1 SE. Note that the multiplicative coefficient of the power function is significantly lowered with bilateral vagotomy, whereas the exponent remains unchanged.

depth of barbiturate anesthesia. This effect of anesthesia can be related to the depressant action of barbiturates on the central drive to intercostal γ-motoneurones[23], that should result in lessened muscle spindle discharge from the extra-diaphragmatic inspiratory muscles and reduced reflex facilitation of their γ-motoneurones. It seems possible that the same mechanisms are involved in the response to vagal afferent signals. Vagal afferent signals could possibly facilitate those supraspinal structures which provide the inspiratory fusimotor drive, whereby the reflex facilitation of the γ-motoneurones of the extra-diaphragmatic inspiratory muscles is ultimately enhanced. This explanation of the effects of both vagotomy and anesthesia on inspiratory muscle activity is supported by the fact that no selective changes in the rate of rise of the moving average EMG of the parasternal muscles are observed if vagotomy is performed or the level of anesthesia is deepened in rabbits with thoracic dorsal rhizotomy[7, 8]. In any case, it appears that vagal afferent signals are not responsible for the different response of the inspiratory muscles to changes in the ventilatory demand: for any given relative change in the rate of rise of diaphragm activity, that of the extradiaphragmatic inspiratory muscles is in fact the same both in intact and vagotomized animals (Fig. 2).

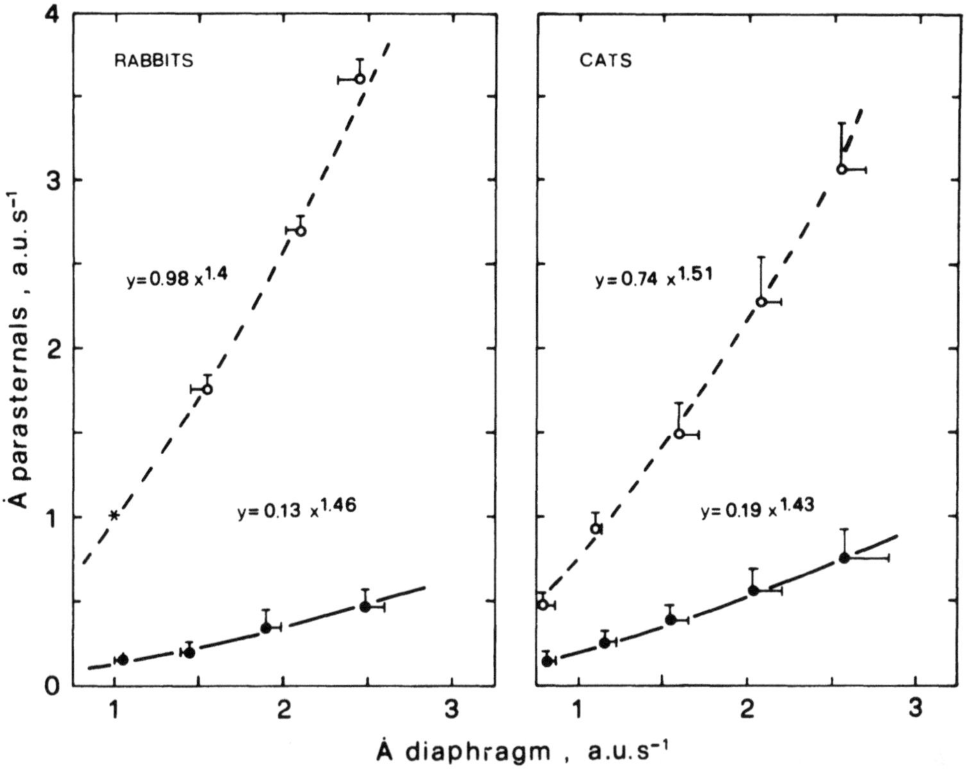

Fig. 3. - The relationship between mean rate of rise of the moving average EMG (A) of the diaphragm and that of the parasternal muscles during rebreathing before (open symbols) and after thoracic dorsal rhizotomy (closed symbols) in rabbits with vagi intact and in cats with bilateral vagotomy. Reference values for each muscle are those obtained during eupneic breathing in the intact animals (asterisk). Data from D'Angelo and Schieppati[7]; bars are 1 SE. Note that the multiplicative coefficient of the power function is markedly reduced, whereas the exponent remains unchanged, independently of intact vagus nerves.

The diaphragm and the extra-diaphragmatic inspiratory muscles have different proprioceptive control. Intercostal γ-motoneurones are driven from supraspinal structures at the inherent inspiratory rhythm and the amplitude of this central modulation increases on enhancing the chemical drive[1, 10, 22, 23, 24], while the majority of the few diaphragmatic muscle spindles are not subjected to rhythmic fusomotor driving[2]. In agreement with these observations, dorsal rhizotomy dramatically reduces the rate of rise of the moving average EMG of the parasternals in rabbits and cats, irrespective of the level of chemical drive, but has no effect on diaphragm activity[7, 8, 20, 21]. In spite of the marked dependence of extra-diaphragmatic inspiratory muscle activity on intact proprioceptive feedback, the greater increase in parasternal muscle than in diaphragm activity on enhancing the ventilatory demand is still present in rabbits and cats with thoracic dorsal rhizotomy (Fig. 3). Since the activity of the inspiratory thoracic muscles after thoracic dorsal rhizotomy necessarily

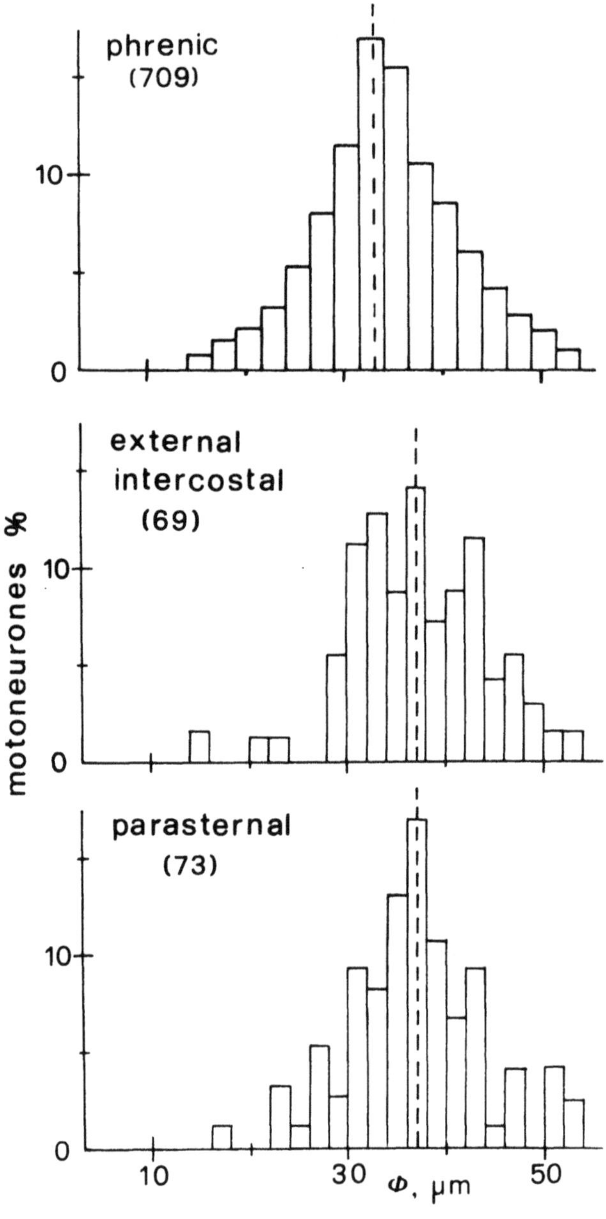

Fig. 4. - Distribution of motoneuron equivalent diameter in the phrenic and inspiratory intercostal motor nuclei of the cat. Total number of sampled cells is given in brackets. The broken line indicates the mean and the mode for cell diameter in each nucleus. Data from Webber et al.[27] and from Larnicol et al.[17]. Transverse cell body dimensions of phrenic motoneurons given by Webber et al.[27] were corrected, taking into account the axial to transverse diameter ratio for phrenic motoneurons given by Webber and Pleschka[26].

reflects only the central command to α-motoneurones, the contribution of the α-γ linkage is to multiply the effects of the central command to α-motoneurones by a constant factor, independent of the respiratory drive. This is in turn consistent with the observation of parallel changes in the central command to γ- and to α-motoneurones of the respiratory thoracic muscles[10, 22].

The persistence of the disproportionate increase of extra-diaphragmatic inspiratory muscle activity relative to that in diaphragm activity with rebreathing after thoracic dorsal rhizotomy indicates that either there are differences in the increase of the central command to inspiratory thoracic and phrenic motoneurones, or the features of the motor nuclei of the inspiratory muscles are substantially different. Data in literature do not favour the hypothesis that mean threshold and threshold distribution differ between the phrenic nucleus and those serving the respiratory thoracic muscles, since, with reference to Henneman's size principle, the mean and the mode of α-motoneurone equivalent diameter coincide in any given motor nucleus and are similar for the various motor nuclei (Fig. 4). The possibility can, however, be entertained that there are differences in connectivity between supraspinal structures and the motor nuclei of individual inspiratory muscles which simulate the effects of differences in mean threshold level of α-motoneurons between the various motor nuclei.

The discussion above strongly suggests that the response of the various inspiratory muscles to changes in the ventilatory demand reflects ultimately some properties of the central inspiratory motor control system. These properties are common to the central inspiratory motor control mechanism of several mammalian species, and it seems reasonable to believe that they are also shared by the inspiratory motor control mechanisms of man. Indeed, the greater increase in extra-diaphragmatic inspiratory muscle activity than in diaphragm activity on enhancing the chemical drive observed in anesthetized animals fits with the greater contribution of rib cage expansion to tidal volume observed in man during CO_2 rebreathing[19]. Moreover, an increase in the ratio of the relative changes in the rate of rise of the moving average EMG of cranial parasternals and of the diaphragm does occur between slow and fast voluntary inspirations in man[5]. Caution must, however, be exercised when extending the results obtained in anesthetized animals to conscious humans. Anesthesia, though light, might in fact have prevented the effects of as yet unknown afferent pathways, acting at central and/or segmental level, from interfering with the central inspiratory command.

References

1. CRITCHLOW V., VON EULER C.: Intercostal muscle spindle activity and its motor control. J. Physiol. (London). 1963. 168: 820-847.
2. CORDA M., VON EULER C., LENNERSTRAND G.: Proprioceptive innervation of the diaphragm. J. Appl. Physiol. (London). 1965. 178: 161-177.

3. D'ANGELO E., MICHELINI S., MISEROCCHI G.: Local motion of the chest wall during passive and active expansion. Respir. Physiol. 1973. 19: 47-59.
4. D'ANGELO E., SANT'AMBROGIO G.: Direct action of the contracting diaphragm on the rib cage in rabbits and dogs. J. Appl. Physiol. 1974. 36: 715-719.
5. D'ANGELO E.: Cranio-caudal rib cage distortion with increasing inspiratory airflow in man. Respir. Physiol. 1981. 454: 215-237.
6. D'ANGELO E.: Inspiratory muscle activity during rebreathing in intact and vagotomized rabbits. Respir. Physiol. 1982. 47: 193-218.
7. D'ANGELO E., SCHIEPPATI M.: Effects of thoracic dorsal rhizotomy or vagotomy on respiratory muscle activity at various levels of chemical drive. Respir. Physiol. 1982. 50: 221-238.
8. D'ANGELO E.: Effects of body temperature, passive limb motion and level of anesthesia on the activity of the inspiratory muscles. Respir. Physiol. 1984. 56: 105-129.
9. D'ANGELO E., GARZANITI N., BELLEMARE F.: Inspiratory muscle activity during unloaded and obstructed rebreathing in dogs. J. Appl. Physiol. 1986. (submitted for publication).
10. EKLUND G., VON EULER C., RUTKOWSKI S.: Spontaneous and reflex activity of intercostal gamma motoneurones. J. Physiol. (London). 1964. 171: 139-163.
11. FARKAS G.A., DECRAMER M., ROCHESTER D.F., DE TROYER A.: Contractile properties of intercostal muscles and their functional significance. J. Appl. Physiol. 1985. 59: 528-535.
12. FARKAS G.A., ROCHESTER D.F.: Contractile characteristics and operating lengths of canine neck inspiratory muscles. J. Appl. Physiol. 1986. 61: 220-226.
13. GANDEVIA S.C., MCKENZIE D.K.: Human diaphragmatic EMG: changes with lung volume and posture during supramaximal phrenic stimulation. J. Appl. Physiol. 1986. 60: 1420-1428.
14. GOLDMAN M.D., GRASSINO A., MEAD J., SEARS T.: Mechanics of the human diaphragm during voluntary contraction: dynamics. J. Appl. Physiol. 1978. 44: 840-848.
15. GRASSINO A., GOLDMAN M.D., MEAD J., SEARS T.: Mechanics of human diaphragm during voluntary contraction: statics. J. Appl. Physiol. 1981. 44: 829-839.
16. GRIMBY G., GOLDMAN M., MEAD J.: Respiratory muscle action inferred from rib cage and abdominal V-P partitioning. J. Appl. Physiol. 1976. 41: 739-751.
17. LARNICOL N., ROSE D., MARLOT D., DURON B.: Anatomical organization of cat intercostal motor nuclei as demonstrated by HRP retrograde labelling. J. Physiol. (Paris). 1982-206.
18. NEWMAN S.L., ROAD J.D., GRASSINO A.: In vivo length and shortening of canine diaphragm with body postural change. J. Appl. Physiol. 1986. 60: 661-669.
19. PENGELLY L.D., TARSHIS A.M., REBUCK A.S.: Contribution of rib cage and abdomen-diaphragm to tidal volume during CO rebreathing. J. Appl. Physiol. 1979. 46: 709-715.
20. SANT'AMBROGIO G., WILSON M.F., FRAZIER D.T.: Somatic afferent activity in reflex regulation of diaphragmatic function in the cat. J. Appl. Physiol. 1962. 17: 829-832.
21. SANT'AMBROGIO G., WIDDICOMBE J.G.: Respiratory reflexes acting on the diaphragm and inspiratory intercostal muscles of the rabbit. J. Physiol. (London). 1965. 180: 766-799.
22. SEARS T.A.: Activity of fusimotor fibres innervating muscles spindles in the intercostal muscles of the cat. Nature (London). 1963. 197: 1013-1014.
23. SEARS T.A.: Efferent discharges in alpha and fusimotor fibres of intercostal nerves of the cat. J. Physiol. (London). 1964, 174: 295-315.
24. SEARS T.A.: Some properties and reflex connexions of respiratory motoneurones of the cat's thoracic spinal cord. J. Physiol. (London). 1964. 175: 386-403.
25. SHARP J.T., GOLDBERG N.B., DRUZ W.S., DANON J.: Relative contribution of rib cage and abdomen to breathing in normal subjects. J. Appl. Physiol. 39: 608-618.
26. WEBBER C.L., PLESCHKA K.: Structural and functional characteristics of individual phrenic motoneurones. Pflugers Arch. 1976. 364: 113-121.
27. WEBBER C.L., WUNSTER R.D., CHUNG J.M.: Cat phrenic nucleus architecture as revealed by horseradish peroxidase mapping. Exp. Brain Res. 1979. 35: 395-406.

A. Grassino - C. Fracchia - C. Rampulla - L. Zocchi
Respiratory Muscles in COPD
Bi & Gi Publishers, Verona - Springer-Verlag, London, 1988

Chest wall and diaphragmatic afferents: their role during external mechanical loading and respiratory muscle ischemia.

Y. Jammes
Laboratoire de Physiologie Hyperbare, GS 15-CNRS, Faculté de Médecine Nord, Marseille, France

SUMMARY: The author presents a briefs examination of the synaptic connections between respiratory muscle afferents and respiratory neurons, followed by a discussion of the afferents' role in eupnea, load-compensatory mechanisms and muscular fatigue.
While this role is an important factor in respiratory response to extreme external mechanical loads (tracheal occlusion at end-expiration; high expiratory threshold loading; positive-pressure breathing), it is not a major component of ventilatory compensation for moderate elastic or resistive loads or internal mechanical loading such as induced bronchospasm.

KEY WORDS: Chest wall; diaphragmatic afferents; loading mechanical; respiratory muscle ischemia; respiratory muscle afferents; respiratory neurons; eupnea; load compensatory mechanisms; muscular fatigue; tracheal occlusion; tracheal end-expiration; high expiratory threshold loading; positive-pressure breathing; ventilatory compensation; bronchospasm.

Skeletal muscles contain both proprioceptors connected to large afferent fibres (groups I and II) and polymodal receptors, which are free neural endings connected to thin fibres (groups III and IV) and sensitive to extracellular fluid composition and also to muscle pressure. Muscle spindles (groups Ia and II fibres) are considered to be length receptors and Golgi tendon organs (group Ib fibres) are force receptors. Numerous studies have been devoted to the role played by groups I and II afferent fibres from the rib cage and abdominal muscles in ventilatory control (see a recent general review by Shannon[19]). However, little information exists on the role played by diaphragmatic receptors and especially on the existence and function of groups III and IV afferent fibres from respiratory muscles. In this paper will be briefly presented first the synaptic connections between respiratory muscle afferents and respiratory neurons, then the role played by these afferents during eupnea and, thirdly, their participation in load-compensatory mechanisms and muscular fatigue will be discussed.

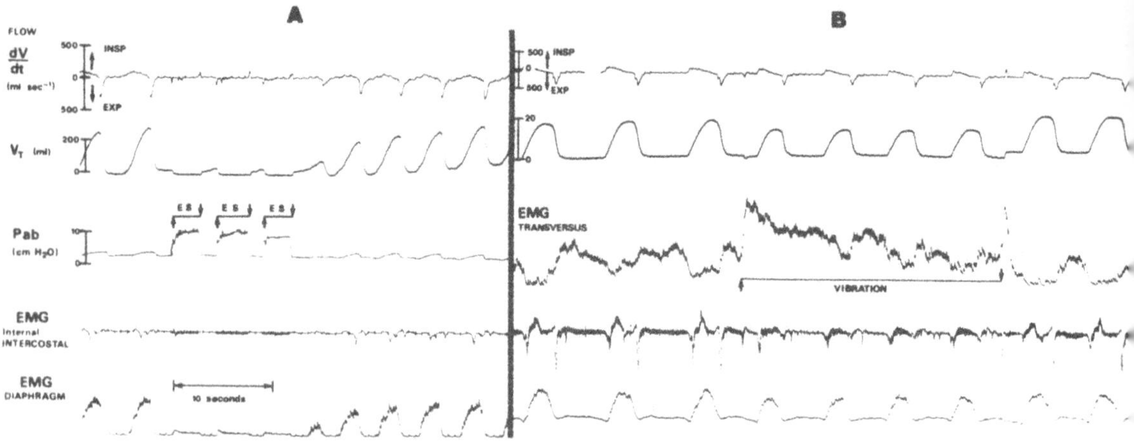

Fig. 1. - Changes in breathing pattern and diaphragmatic activity induced by rhythmic tetanic contractions of the transversus muscle (electrical stimulation: ES) (part A) or during the tonic vibratory response contraction produced by high-frequency mechanical vibrations (part B).

1. Neural connections between respiratory muscle afferents and respiratory neurons.

This neurophysiological basis is necessary to understand the role played by the different respiratory muscle afferents in the control of breathing.

There are spinal segmental reflexes between intercostal muscles involving groups I and II muscular afferents. The stimulation of afferents from muscle spindles (groups Ia and II) activates the ispilateral alpha motoneurons of the same and adjacent spinal segments. High frequency mechanical vibrations applied on an intercostal space activate muscle mechanoreceptors, including spindle endings, and induce a reflex tonic vibratory response (TVR) recorded on the EMG of the corresponding intercostal muscle[8]. Conversely, tendon organ afferents from rib cage muscles exert an inhibitory effect on their homonymous alpha motoneurons of the same segment. There are also proprioceptor reflexes in the abdominal muscles[3]. Figure 1A shows that TVR can be also recorded on EMG of tranversus muscle when mechanical vibrations are applied on the linea alba[9]. There is little information on the existence of phrenic reflexes. However, in cats or rabbits the stimulation of diaphragmatic tendon organs inhibits the phrenic motor output[6]. Because of the paucity of spindles in the diaphragm, reflex facilitation of phrenic motoneurons during diaphragm stretching seems to occur in very rare circumstances[19].

On the other hand, intersegmental effects of intercostal muscle spindles and Golgi tendon organs on phrenic motoneurons are well documented. The stimulation of midthoracic segments (T3-T8) reduces the phrenic motor discharge, whereas the stimulation of afferent fibres from lower thoracic intercostal and abdominal muscles facilitates the phrenic activity in vagotomized cats[19]. However, electrically-induced contractions in transversus muscles (Fig. 1A) or TVR of these muscles (Fig. 1B) are associated with a marked reduction in phrenic motor discharge in vagotomized dogs[9].

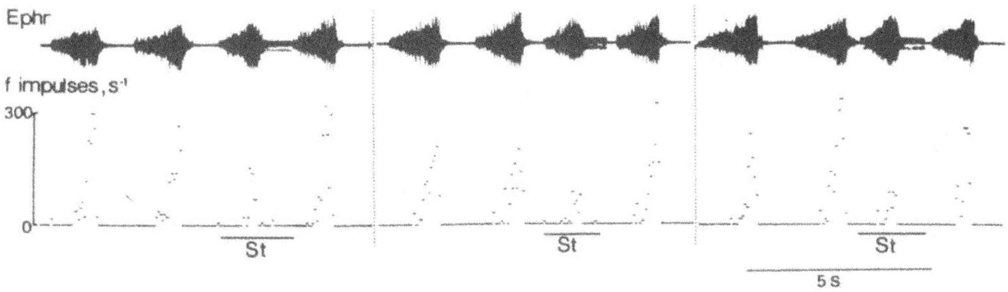

Fig. 2. - Electrical stimulation (St) of the central end of left cut phrenic nerve reduces the impulse frequency (F impulses) of phrenic motoneurons and initiates the inspiratory activity when beginning early in the expiratory period.

The supraspinal effects of intercostal muscle proprioceptors on respiratory control are predominantly inhibition of both inspiratory neurons in the medullary dorsal and ventral respiratory groups, and expiratory neurons in the nucleus retroambigualis. This occurs with stimulation of afferent fibres in external and internal intercostal nerves and seems to involve mostly group Ib (Golgi tendon organs) and perhaps group II, but not group Ia, fibres (primary muscle spindle endings). All these effects result in increased respiratory frequency. They are not so evident during the stimulation of abdominal muscle afferents (Fig. 1). Recently, we have also shown that peripheral electrical stimulation of the phrenic nerve increased the respiratory frequency and this was due to a prominent shortening of the inspiratory duration[11]. Phrenic nerve stimulation can also trigger the inspiratory activity when it begins early during the expiratory phase (Fig. 2). This ventilatory response to the stimulation of phrenic afferents requires the integrity of nerve conduction in both large (group I) and small phrenic fibres (groups III and IV).

2. Role of respiratory muscle afferents during eupnea

Data concerning the role played by rib cage afferents is contradictory. Dorsal root sectioning from T1 to T12 or spinal section at C8 level elicits either changes in both tidal volume and respiratory frequency[4, 5, 17], or isolated changes in the tidal volume[12], or no effect[20]. As suggested by Shannon[19], these different effects are probably the result of the interruption of thoracic reflexes which directly or indirectly affect the respiratory pattern. This is because the central integration of these spinal reflexes depends on the depth and the type of anesthesia[12] in the different studies.

In addition, afferent inputs from diaphragmatic proprioceptors (group Ib) influence the spontaneous phrenic motor activity and also the eupneic respiratory rhythm in cats[11]. This influence was shown during cold block at 6 C of large phrenic fibres, but did not occur during selective block of conduction in groups III and IV phrenic fibres by procaine. These results are not contradictory because large afferent phrenic fibres (Golgi tendon organs) display a spontaneous high-frequency

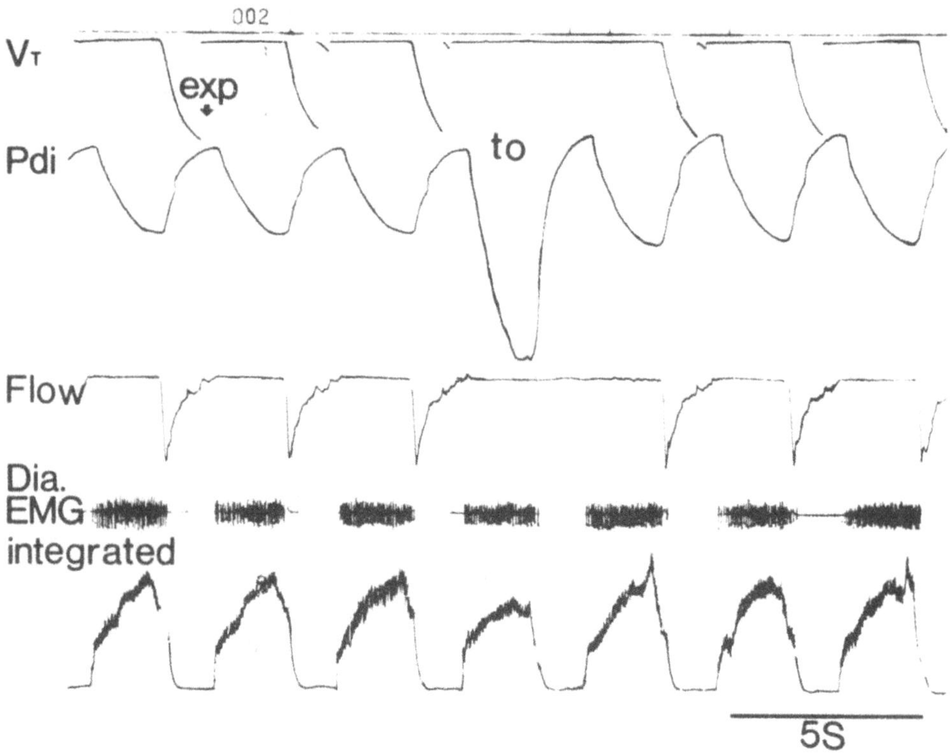

Fig. 3. - Single breath tracheal occlusion (TO) at end-expiration in vagotomized and spinal (C8 section) cat reduces the amplitude of integrated diaphragmatic EMG but does not significantly alter the ventilatory timing.

discharge (30 to 40 impulses s^{-1}) in phase with the diaphragmatic contraction, but the spontaneous activity of small phrenic fibres is irregular and sparse with a low frequency (less than 14 impulses. s^{-1}) [7, 11].

Thus, thoracic nerve afferent information does not seem to play a major role in the control of breathing pattern during eupnea but, at least in cats, proprioceptive inputs from the diaphragm may participate in this control.

3. Role of respiratory muscle afferents during mechanical ventilatory loading

First, external mechanical loading such as chest compression, tracheal occlusion, resistive or elastic loads must be distinguished from internal loading such as drug-induced bronchospasm or passive inflation or deflation of the lungs. In each circumstance, single-breath loading, which does not involve changes in chemoreflex drive, must also be differentiated from multi-breath (or steady state) ventilatory loading. In the latter situation, data analysis requires constant arterial blood gases. In all cases, load-induced changes in respiratory muscle activity must be considered first in intact, then in vagotomized animal preparations, in order to suppress the effects due to vagal stimulation.

- *External mechanical loading*

Chest compression or tracheal occlusion (TO) at end-expiration (infinite inspiratory load) decreases the thoracic gas volume. In vagotomized animals with intact spinal cord, these loads prolong the diaphragmatic contraction for the first loaded breath[12], and this corresponds to prolonged firing duration to 50% of medullary inspiratory neurons in dorsal and ventral respiratory groups[18]. These effects disappear after cervical plus thoracic dorsal rhizotomies, but persist with either cervical or thoracic dorsal roots intact. This is not observed in all experimental conditions. Thus, in vagotomized and spinal (C8 spinal section) cats under chloralose-urethan anesthesia, TO at end-expiration induced no change in the duration of phrenic nerve activity but, as shown in Figure 3, this markedly reduces the amplitude of integrated diaphragmatic EMG[11]. TO at end-expiration increases the discharge of both spindle endings and tendon organs in external intercostal and parasternal muscles[12, 19] and of Golgi tendon organs in the diaphragm[11]. The stimulation of diaphragmatic Golgi tendon organs could be responsible for the inhibition of phrenic motoneurons, as observed during TO in spinal and vagotomized cats, but excitation of spindle endings, very numerous in rib cage muscles, could account for the augmentation of medullary inspiratory neuron activity[19]. Actually, both segmental and suprasegmental spinal reflexes from the rib cage and diaphragm seem to participate in the response to TO at end-expiration. Sustained compression of large areas of the chest in vagotomized cats induces various changes in respiratory frequency with in most cases, shortening of inspiratory duration[19]. Conversely, multibreath TO at end-expiration prolongs the activity of external intercostal muscles and the diaphragm and can induce tonic contractions in these muscles in anesthetized rabbits (personal observations). These tonic activities rise in proportion to decreased lung volume with TO and do not involve respiratory muscle afferents because they are no longer induced after bivagotomy.

Reciprocal effects occur during expiratory loading, which increases the volume of the chest wall. TO at end-inspiration increases the internal intercostal muscle activity for the first loaded breath but does not change the discharge of medullary inspiratory neurons[19]. Multibreath TO at end-inspiration does not induce tonic contraction in respiratory muscles (internal intercostal and abdominal muscles) and increases the diaphragmatic contraction. This also occurs during high expiratory threshold loading (ETL) or positive-pressure breathing when the chemical drive for breathing is maintained constant under cardiopulmonary bypass[9]. Evoked expiratory activities and enhanced diaphragmatic efforts need the integrity of pulmonary vagal afferents to be maximal[9, 16]. The most important consequence of evoked contraction in expiratory muscles with ETL breathing is that the central integration of vagal information from the lung and of chemosensory reflexes (response

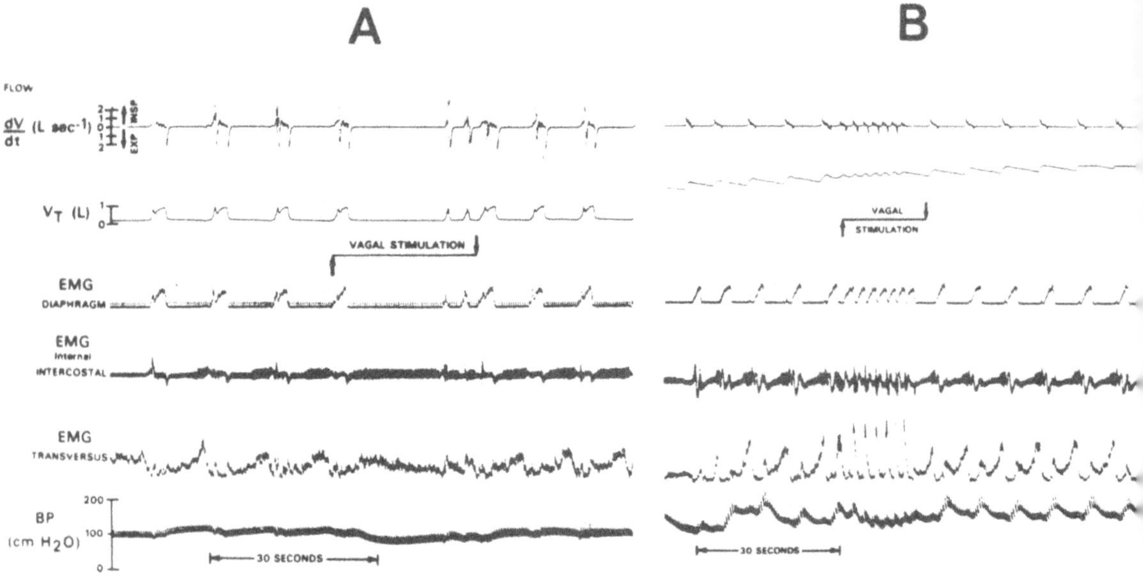

Fig. 4. - Response to vagal stimulation with electrical shocks during unloaded breathing (A), then after 10mn of ETL breathing in a dog under extracorporeal circulation. The inspiratory inhibition associated with evoked tonic contraction in expiratory muscles (transversus) is replaced by a tachypnea.

to hyper-capnia) is progressively impeded and even disappears during prolonged loading[9, 10]. This seems to be the consequence of a competition between increased afferent discharge from contracting respiratory muscles and vagal or chemosensory inputs. Figure 4 gives an example of the suppression of the inhibitory response to vagal stimulation after 10mn of ETL breathing in a dog under extracorporeal circulation. This phenomenon has not yet been studied during inspiratory loading.

Elastic loading increases the respiratory frequency in vagotomized cats and dogs but this does not occur when the chemical drive is held constant[19]. Thus, respiratory muscle afferents do not participate in this response. Single breath inspiratory or expiratory resistive loading in vagotomized cats cannot stimulate the chemoreflex drive; in this situation, the inspiratory duration and the respiratory frequency do not change, but the magnitude of integrated diaphragmatic EMG activity increases with I or E loading[14] (Fig. 5). This effect is suppressed by spinal cord section at C8 level, revealing that facilitatory influences from respiratory muscles participate in the respiratory compensation for external resistive loads but that their effects are counterbalanced by inhibitory inputs from pulmonary vagal receptors.

- *Internal mechanical loading*

Passive lung inflation in vagotomized animals exerts various and often opposite effects on the respiratory muscle activity. Duron[4] reports an inhibition of diaphragm, external intercostal and interchondral muscles in cats, attributed to increased activity of muscle spindle endings and tendon organs in antagonistic muscles (stretch of internal intercostal muscles during chest distension). However, Bainton

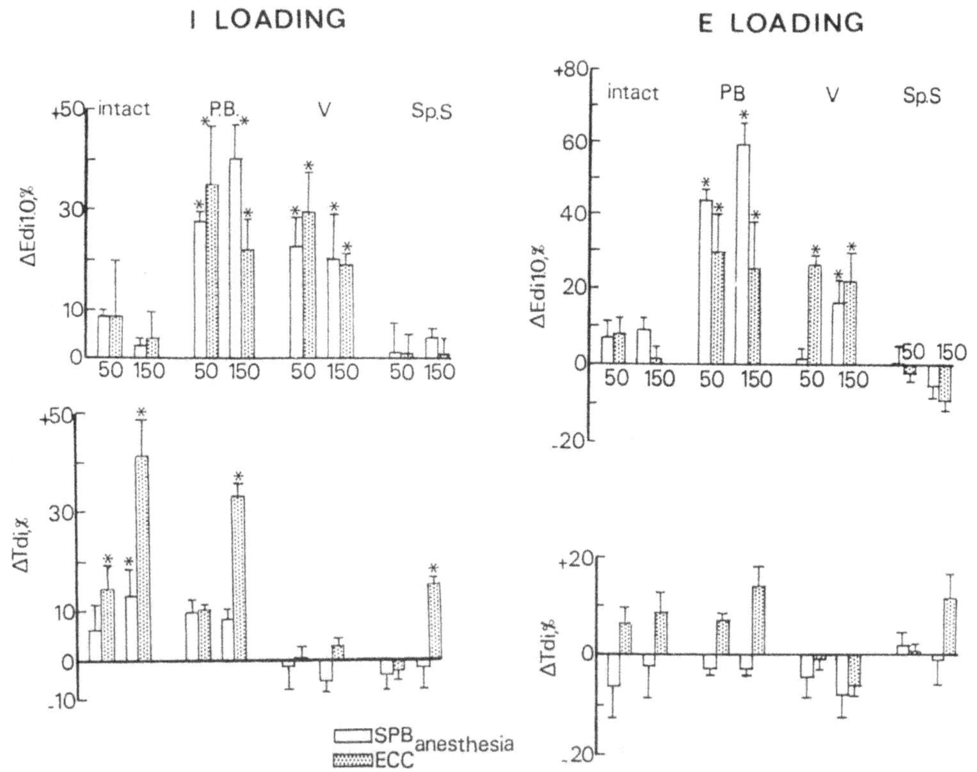

Fig. 5. - Changes in the magnitude of integrated diaphragmatic EMG (Edi) and in duration of diaphragmatic activity (Tdi) induced by external resistive inspiratory or expiratory loads (50 and 150 cm $H_2O.l^{-1}.s^{-1}$). These effects are studied in intact cats, after selective procaine block (PB) of small vagal fibres, vagotomy (V) or spinal section at C8 level (SpS).

et al.[1] reported the opposite effects in the same species, and another study by Bianchi and Barillot[2] has shown that passive expansion of the chest wall does not induce significant changes in central respiratory activity. On the other hand, lung deflation elicits a marked increase in respiratory frequency in all animal species, associated with evoked tonic contractions in inspiratory muscles in rabbits (personal observation). However, these effects disappear after bivagotomy and seem only to involve the stimulation of lung deflation receptors (type J vagal receptors and bronchial C-fibres).

Acute bronchospam induced by i.v. injection of histamine, carbachol or phenyldiguanide produces tachypnea plus marked tonic inspiratory activities in anesthetized rabbits (personal observations). All these effects persist after atropine block of drug-induced bronchoconstriction, but disappear after vagotomy or selective block of conduction in small vagal fibres and thus do not result from reflex pathways from respiratory muscles.

4. Role of respiratory muscle afferents during muscular ischemia and muscle fatigue

In skeletal muscles groups III and IV fibres are activated during periods of muscular ischemia and by accompanying extracellular changes, such as intramuscular accumulation of lactic acid and hyperosmolarity[13]. As yet, the existence of these afferent fibres from the respiratory muscles has been demonstrated only for the diaphragm[7, 11]. The selective electrical stimulation of thin phrenic afferent fibres is able to alter the central respiratory timing. Changes in timing are also observed during the response to breathing against a fatiguing load or in a state of reduced respiratory blood flow (as in shock), which induces tachypnea initially, followed by bradypnea and respiratory arrest[15]. Because this response was not affected by vagotomy or cross perfusion of the head (eliminating chemoreceptor afferents), it is tempting to speculate that the activation of groups III and IV afferent phrenic fibres may play a role in the observed changes in respiratory timing during shock.

In addition, the spontaneous activity of phasic afferent phrenic fibres which exhibit a peak of discharge during diaphragmatic contraction (Golgi tendon organs) is decreased from control during diaphragmatic ischemia, lactic acid or hypertonic NaCl injection[7]. This suggests that, at least for the diaphragm, changes in extracellular fluid composition may uncouple the discharge of diaphragmatic mechanoreceptors from muscle tension, although no existing studies have dealt with a similar point concerning proprioceptors in skeletal muscles. As detailed above, some animal studies have shown that these diaphragmatic receptors participate in ventilatory control[6, 11] and, as a consequence, changes in breathing pattern observed during diaphragm fatigue may result from decreased influence of diaphragmatic proprioceptors and/or from increased discharge of groups III and IV phrenic fibres.

In conclusion, numerous observations concur in demonstrating that respiratory muscle afferents contribute to the ventilatory response to extreme external mechanical loads (such as TO at end-expiration or severe ETL and positive-pressure breathing), but do not play a major role in the respiratory compensation for moderate elastic or resistive loads or internal mechanical loading (bronchospasm).

References

1. BAINTON C.R., KIRKWOOD P.A. SEARS T.A. On the transmission of the stimulating effects of carbon dioxide to the muscles of respiration. J. Physiol. (London). 1978, 280: 249-272.

2. BIANCHI A.L. BARILLOT J.C.. Activity of medullary respiratory neurons during reflexes from the lungs in cats. Respir. Physiol., 1975. 25: 335-352.

3. BISHOP B. Reflex control of abdominal muscles during positive-pressure breathing. J. Appl. Physiol. 1964. 19: 224-233.
4. DURON B. Postural and ventilatory functions of intercostal muscles. Acta Neurobiol. Exp. 1973. 33: 355-380.
5. GAUTIER H. Respiratory responses of the anesthetized rabbit to vagotomy and thoracic dorsal rhizotomy. Respir. Physiol., 1973. 17: 238=247.
6. GILL P.C. KUNO M., Excitatory and inhibitory actions on phrenic motoneurones. J. Physiol. (London), 1963. 168: 274-289.
7. GRAHAM R., JAMMES Y., DELPIERRE S., GRIMAUD C., ROUSSOS C.. The effects of ischemia, lactic acid and hypertonic sodium chloride on phrenic afferent discharge during spontaneous diaphragmatic contraction. Neuroscience Letters, 1986. 67: 257-262.
8. HOMMA I., EKLUND G., HAGBARTH K.E.. Respiration in man affected by TVR contractions elicited in inspiratory and expiratory intercostal muscles. Respir. Physiol., 1978. 35: 335-348.
9. JAMMES Y., BYE P.T.P., PARDY R.L., KATSARDIS C., ESAU S., ROUSSOS C. Expiratory threshold load under extracorporeal circulation: effects of vagal afferents. J. Appl. Physiol., 1983. 55: 307-315.
10. JAMMES Y., BYE P.T.P., PARDY R.L., ROUSSOS C.. Vagal feedback with expiratory threshold load under extracorporeal circulation. J. Appl. Physiol., 1983. 55: 316-322.
11. JAMMES Y., BUCHLER B., DELPIERRE S., RASIDAKIS A., GRIMAUD C., ROUSSOS C.. Phrenic afferents and their role in inspiratory control. J. Appl. Physiol., 1986. 60: 854-860.
12. JAMMES Y., MATHIOT M.J., DELPIERRE S., GRIMAUD C. Role of vagal and spinal sensory pathways on eupneic diaphragmatic activity. J. Appl. Physiol., 1986. 60: 479-485.
13: KAUFMAN, M.L., RYBICKI K.J., WALDROPS T.G., ORDWAY G.A.. Effects of ischemia on responses of group III and IV afferents to contraction. J. Appl. Physiol., 1984. 57: 644-650.
14. MATHIOT M.J., JAMMES Y., GRIMAUD C. Role of vagal and spinal sensory pathways in diaphragmatic response loading. Neuroscience Letters (in press).
15. ROUSSOS, C. Ventilatory muscle fatigue governs breathing frequency. Clin. Respir. Physiol., 1984. 20: 445]451.
16. RUSSEL J.A., BISHOP B.. Vagal afferents essential for abdominal muscle activity during lung inflation in cats. J. Appl. Physiol., 1976. 41: 310-315.
17. SHANNON R. Effects of thoracic dorsal rhizotomies on the respiratory pattern in anesthetized cats. J. Appl. Physiol., 1977. 43: 20-26.
18. SHANNON R., SHEAR W.T., MERCAK A.R., BOSLER D.C., LINDSEY B.G.. Non-vagal reflex effects on medullary inspiratory neurons during inspiratory loading. Respir. Physiol., 1985. 60: 193-204.
19. SHANNON R. Reflexes from respiratory muscles and costo-vertebral joints. In: Handbook of Physiology. The respiratory system II. 1986. 431-447.
20. SPECK D.F. WEBBER C.L.. Thoracic dorsal rhizotomy in the anesthetized cat: maintenance of eupneic breathing. Respir. Physiol., 38: 347-357.

A. GRASSINO - C. FRACCHIA - C. RAMPULLA - L. ZOCCHI
Respiratory Muscles in COPD
Bi & Gi Publishers, Verona - Springer-Verlag, London, 1988

The oxygen cost of respiratory and non-respiratory muscles

S. SANCI[1], S. ROMANO[1], S. FIELD[2], V. BELLIA[1], G. BONSIGNORE[1], A. GRASSINO[2]

[1] *Respiratory Physiopathology Department, National Research Council, Palermo, Italy*
[2] *Meakins-Christie Laboratories, McGill University, and Notre-Dame Hospital, University of Montreal, Quebec, Canada*

SUMMARY: From the results obtained in these two sets of experiments we conclude: a) a considerable cause of variability in $\dot{V}O_2$ resp must be the recruitment of large postural muscles, particularly during unlearned tasks as targeting patterns. Since the weight of those muscles is considerably larger than that of the respiratory muscles, even a milder activity may considerably influence the $\dot{V}O_2$. The use of postural muscles can be monitored and minimized by training or choosing adequate body position. b) The pressure-time index of the diaphragm, where pressure is expressed as a fraction of actual Pdimax corrected for prevailing velocity of shortening, constitutes a very good mechanical index of $\dot{V}O_2$ resp over a wide variety of breathing patterns in which T_I/T_T changes. This relationship will weaken if non-respiratory muscles are active.
KEY WORDS: Oxygen cost; respiratory muscle; non-respiratory muscle; large postural muscle; postural muscle; diaphragm; breathing patterns; mechanical indexes; $\dot{V}O_2$ resp; resistive breathing; hyperventilation.

Introduction

The oxygen consumed by the respiratory muscles in normal subjects during quiet breathing ($\dot{V}O_2$ resp) is a small proportion of the total oxygen requirement ($\dot{V}O_2$ tot). Although estimates of the magnitude of $\dot{V}O_2$ resp vary considerably[14, 21], it is believed to be about 1-2% of $\dot{V}O_2$ tot[25]. The measurement of the oxygen cost of breathing in humans is difficult because it cannot be measured directly. The technique consists of measuring first the $\dot{V}O_2$ tot at rest and then increasing ventilation to different levels and calculating the new $\dot{V}O_2$. The increase over resting level is assumed to be $\dot{V}O_2$ caused by respiratory muscles. The unproved assumption is that the increased oxygen uptake will represent only the metabolic cost of the increase in the respiratory muscles' work. This is an assumption we tested and disproved, and whose implications are discussed below. As ventilation increases, the $\dot{V}O_2$ resp increases hyperbolically. Figure 1 shows the relationship between $\dot{V}O_2$ resp and $\dot{V}_E$. These data were obtained in 1960 by Milic-Emili et al.[21], who measured the energy cost of breathing in four normal supine subjects during hyperpnea induced by dead space breathing.

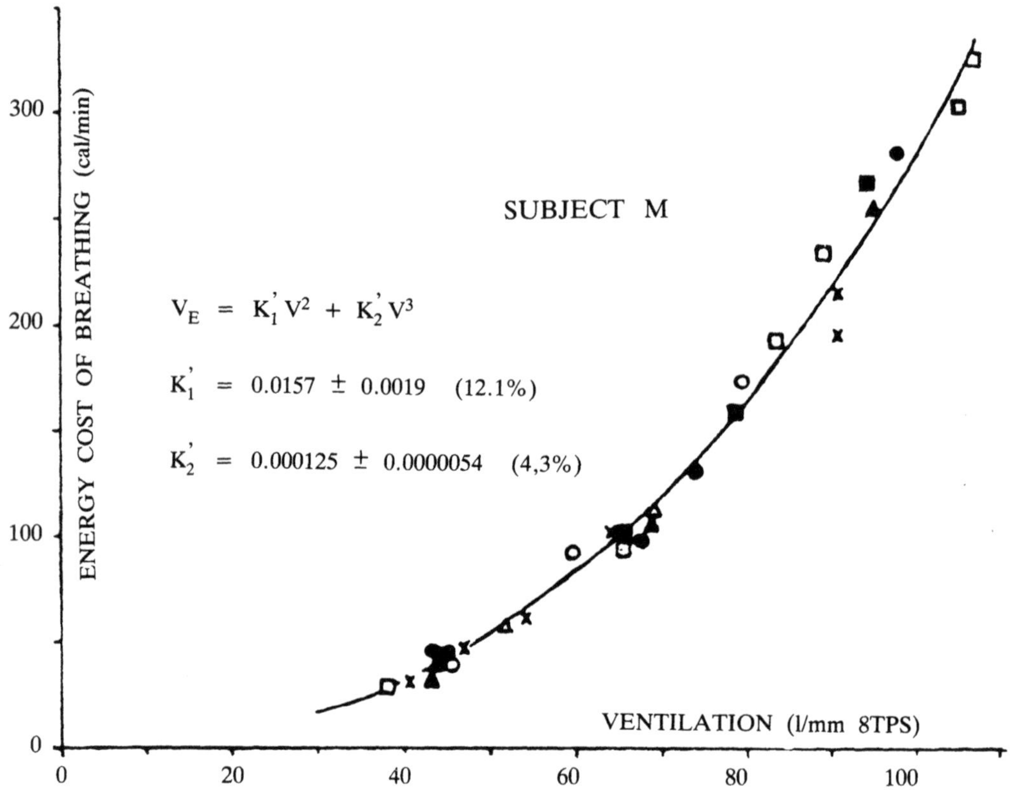

Fig. 1. - Ordinate: energy cost of breathing in cal/min, calculated by converting to calories the O_2 consumption of respiratory muscles. Abscissa: pulmonary ventilation in L/min. [21]

However, it is well known that the slope of the hyperbola is variable among subjects, and that there is a large variation calculated at isoventilation among different investigators[4, 7, 10, 11, 14, 20] in the reported values the of O_2 cost of breathing.

Figure 2 shows values of $\dot{V}O_2$ resp at isoventilation (60 l/min) obtained by different investigators. It indicates that there is a wide variability in the estimates of $\dot{V}O_2$ resp ranging from 30 to 230 ml O_2min^{-1}. This large spread can be explained either by physiological variation between subjects or by variations related to the experimental methods for $\dot{V}O_2$ measurements.

Lower values of $\dot{V}O_2$ resp have been usually obtained during hyperventilation induced by dead space in trained subjects[21], whereas higher values have been obtained during voluntary hyperventilation[20] or inspiratory resistive breathing[19]. One of the physiological factors that may account for such a difference is the active recruitment of the accessory and postural muscles, which, while not directly contributing either to inflating the lungs or to generating transdiaphragmatic pressure, will consume oxygen.

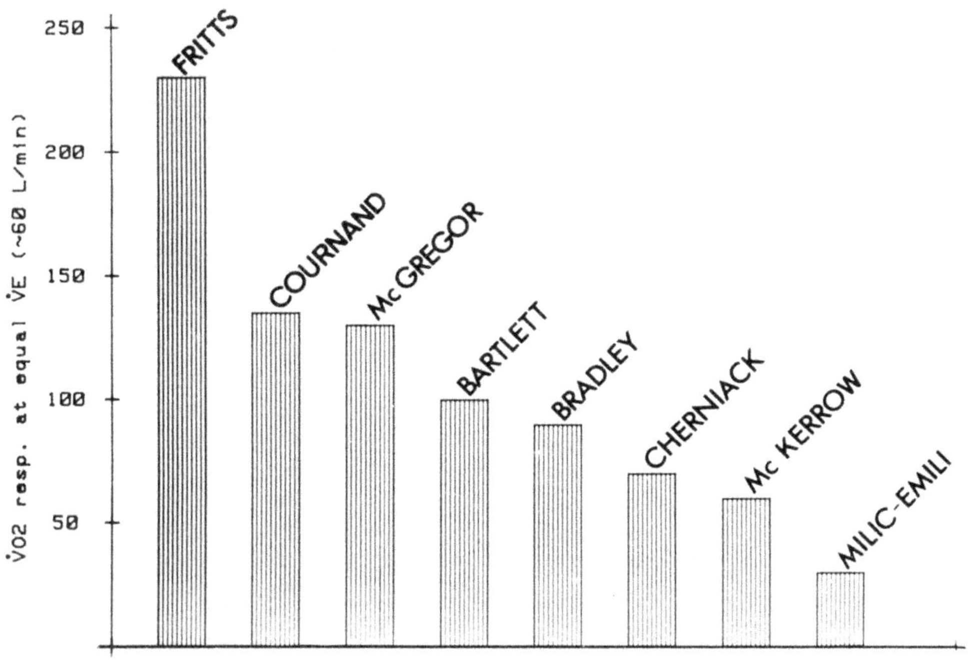

Fig. 2. - Oxygen cost of breathing at equal (about 60 1/min) ventilation. Note the marked difference in O_2 consumption obtained by different investigators.

Mechanical indexes of VO² resp

Respiratory oxygen consumption is often related to the external work of breathing as calculated by the Campbell method (pressure-volume), and their "efficiency" is derived[8, 9]. Efficiency is reported to vary from 1% to 25% depending on the author, a wide spread unique to the respiratory muscles and often attributed to a variable "coupling" of the chest wall muscles.

Following Hill's studies[17], Mommaerts[22] established that the energy cost of a skeletal muscle contraction depends on: U = A + L + W + f (P,t), where the energy cost of a contraction (U) is equal to the sum of energy due to muscle activation (A), shortening (L), work performed (W) and a function of the pressure developed and its duration. Whether any one of these parameters alone or any combination of them can account for the total O_2 requirement of the respiratory muscles is not clear. For example, work cannot account for changes in the time of a contraction, for example when T_I/T_T is changing, as in CO_2-induced hyperpnea or in exercise.

However, a parameter that reflects both the tension developed and its duration is expected to reflect a larger proportion of the $\dot{V}O_2$ resp than the work of breathing (W). Such a parameter is the time integral of Pdi. McGregor and Becklake[19] found that respiratory muscle force was a better index than W to

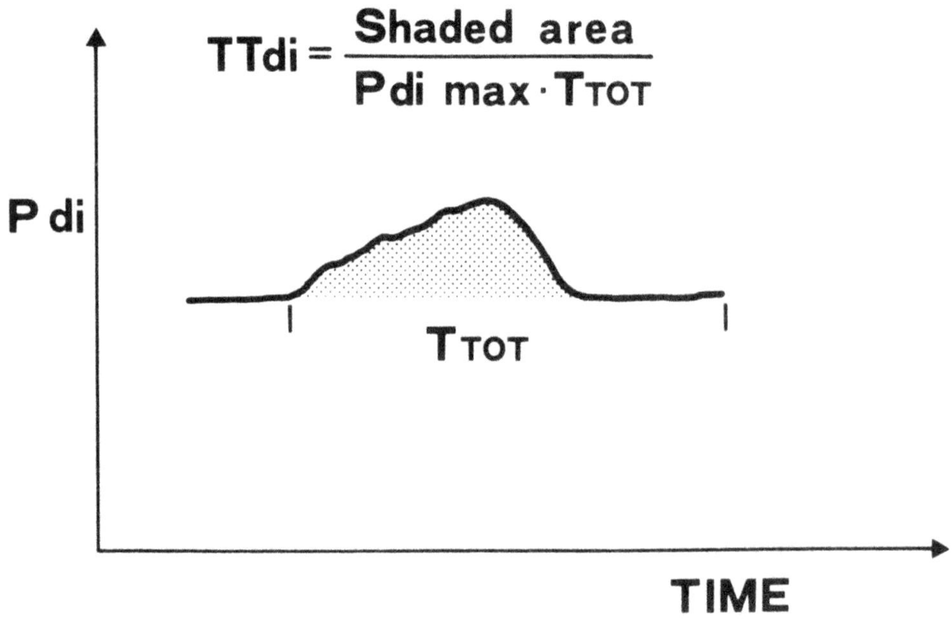

Fig. 3. - The shaded area delimited by the Pdi trace was measured and expressed as a fraction of Pdimax and total breath cycle duration (T_T). TTdi = Pdi.dt: Pdimax. T_T.

estimate the $\dot{V}O_2$ resp. Moreover, Rochester and Bettini[24] found a strong linear relationship between the pleural pressure time index and the O_2 consumption of the diaphragm in dogs both during loaded and unloaded CO_2-induced hyperventilation.

Methods and results: *Tension time and work*

In the following experiments we describe the relationship between $\dot{V}O_2$ resp - work of breathing and tension time in human subjects under two experimental conditions: 1) voluntarily breathing against an inspiratory resistance[13]; 2) during hyperventilation induced by dead space. Subjects were highly trained to perform such experiments and were placed in a comfortable position so as to minimize recruitment of non-respiratory muscles.

Figure 3 shows how the TTI of the diaphragm is usually measured[5]. It is the area delimited by the Pdi trace (shaded area), expressed as a fraction of the product of Pdimax and total breath cycle duration (T_T).

$$TTdi = Pdi.dt \div Pdimax.T_T$$

The oxygen consumption was measured by the closed circuit method and was calculated as the slope of the end expiratory volume points. The difference between the O_2 consumption observed during augmented breathing and resting ventilation was the value of $\dot{V}O_2$ resp.

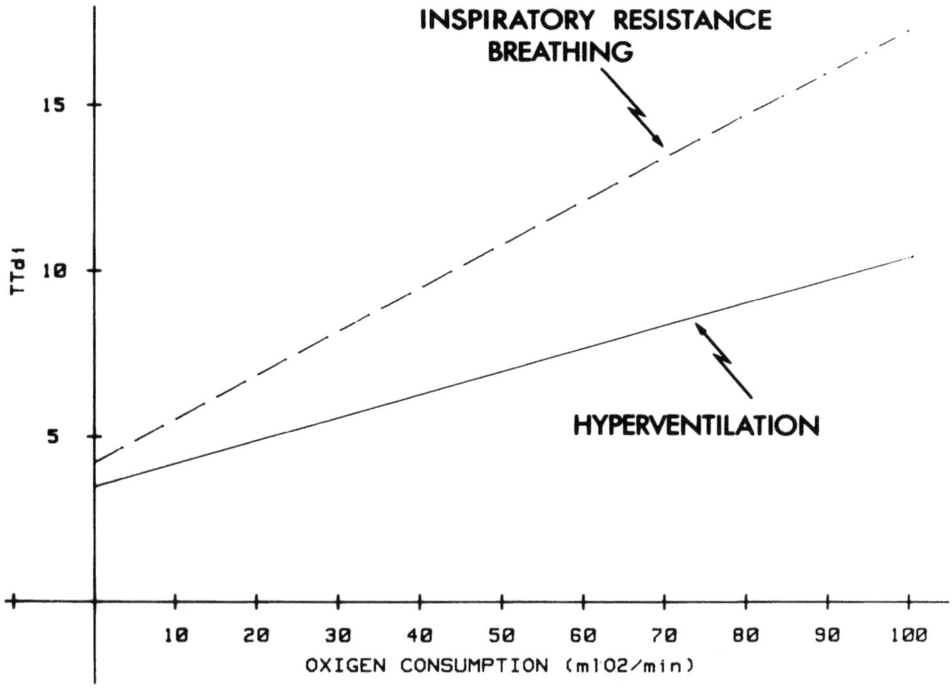

Fig. 4. - Regression lines between the TTdi and $\dot{V}O_2$ resp. The broken line represents the results obtained during resistive breathing, the continuous line, the results obtained during hyperventilation. The slopes were statistically different ($p < 0.001$).

We studied four normal male subjects who performed a total of 33 runs breathing against an inspiratory resistance and a total of 27 runs during hyperventilation induced by dead space. All runs had a TTdi below the fatigue threshold.

A strong linear relationship between the TTdi and $\dot{V}O_2$ was found irrespective of the experimental conditions adopted, whereas the work of breathing correlated significantly to $\dot{V}O_2$ resp only during hyperventilation.

However, the slopes of the regression lines for dead space and inspiratory resistive breathing obtained for the TTdi and $\dot{V}O_2$ resp were statistically different, as shown in Figure 4. For a given value of TTdi the $\dot{V}O_2$ resp was higher during hyperpnea than during resistive breathing (i.e. more oxygen was consumed for a given TTdi during hyperpnea).

Several factors can account for the observed difference. First of them could be the Fenn effect[12], i.e. when a muscle is performing work, more energy is required than if the muscle has an isometric contraction. During hyperpnea the external work performed by the respiratory muscles increases with ventilation, whereas during high inspiratory resistance breathing small displacement takes place and, despite large

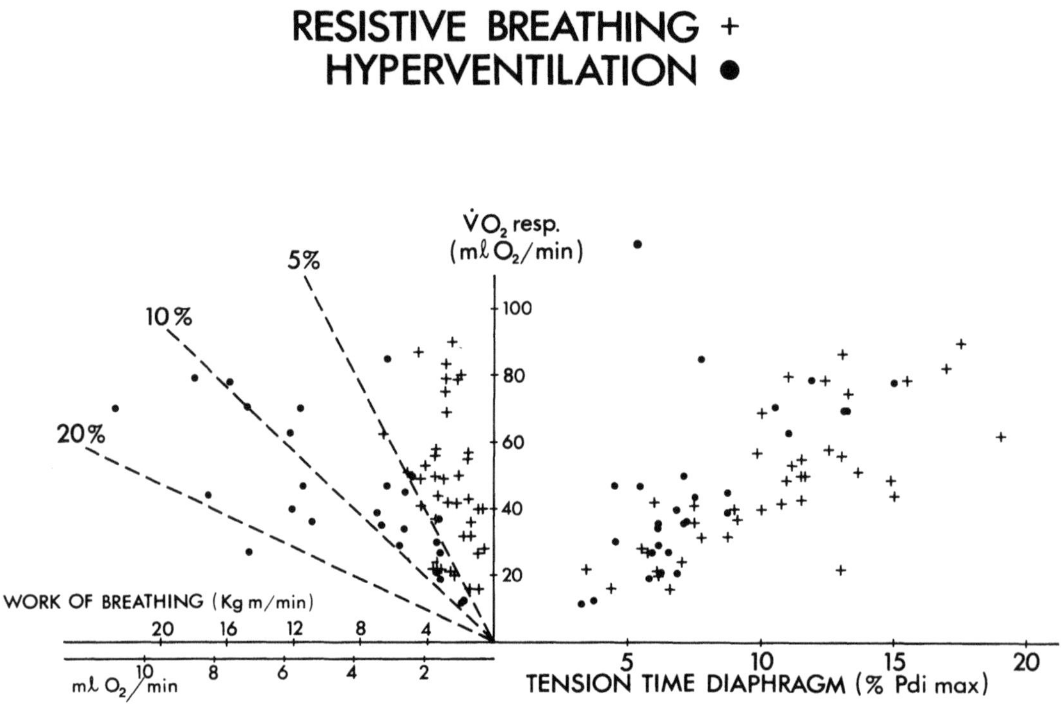

Fig. 5. - Relationship between $\dot{V}O_2$ resp, work of breathing (left panel), and TTdi (right panel). Broken lines in left panel are respiratory muscle efficiency isopleths. Different symbols represent data points of different experimental conditions as follows: + resistive breathing; o hyperventilation. The hyperventilation TTdi values were expressed as a fraction of actual Pdimax corrected for V_T/T_I.

pressure swings the mechanical work is smaller. T_I/T_T is considerably larger in resistive breathing.

Secondly, according to the force-velocity relationship[2], the maximum inspiratory pressure that a muscle can generate at a given lung volume decreases with increasing flow rates. The mean inspiratory flow will indirectly reflect the velocity of shortening. During hyperventilation the subjects developed V_T/T_I values that were five times greater than during resistive breathing. The expected result is a decrease in Pdimax with increasing V_T/T_I. Therefore, the TTdi should be corrected according to the prevailing Pdimax at any given V_T/T_I. The reduction in Pdimax at high V_T/T_I will increase the numerical value of TTdi. Applying this correction, we found a unique relationship between TTdi and VO_2 resp both during hyperventilation and resistive breathing. The results are shown in Figure 5, right panel, where different symbols represent different experimental conditions. According to our results the TTdi is a reliable and relatively easy to measure mechanical index of the oxygen cost of breathing and it may predict the VO_2 resp under different experimental conditions, including resistive breathing and CO_2-induced hyperpnea.

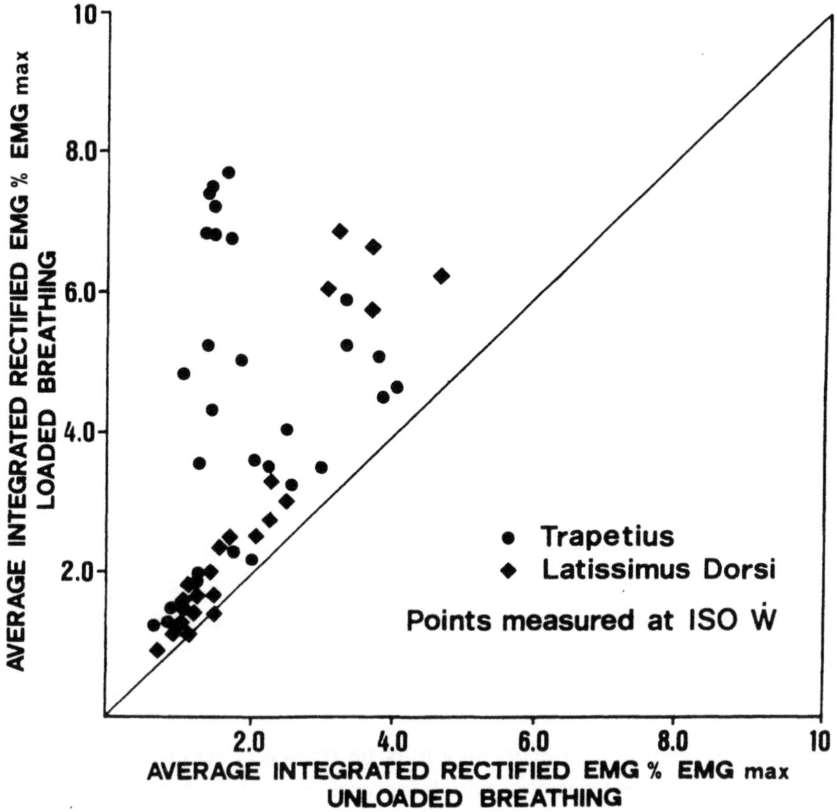

Fig. 6. - Average integrated rectified EMG activity expressed as % of the EMG max. T_I/T_T. Vertical axes: values obtained during loaded breathing; horizontal axes: values obtained during unloaded breathing. Each point was obtained at isowork values. Different symbols represent different muscles as follows: ● trapetius; ■ latissimus dorsi. It is shown that at isowork the EMG activity was higher during loaded than unloaded breathing. Continuous line: identity line.

Efficiency of breathing

Respiratory muscle efficiency is the ratio of the work of breathing performed to the oxygen consumed by the respiratory muscles. Reid and Johnson[23] studied the mechanical efficiency of the diaphragm by measuring the mechanical work of the diaphragm and the O_2 consumption at different inspiratory resistive loads. They found that the mechanical efficiency of the diaphragm averaged 23% and was constant as work loads were increased. Their results confirm that an efficiency value of 25% seems to be reasonable for the diaphragm, as already observed for other skeletal muscles. However, respiratory muscle efficiency has proved to be a parameter with a wide variation among different experimental conditions reported. Measurements of efficiency range from 1 to 25%, depending upon the type of subject studied and the methods used to increase both the $\dot{V}O_2$ resp and W. Lower efficiency values have usually been obtained during loaded breathing[13]. In our studies, efficiency values ranged from 1 to 25% and were lower during resistive breathing (Fig. 5, left panel) than during hyperpnea. Several reasons have been proposed to explain such a large dispersion.

The proposed causes include the type of respiratory loads on chest wall coupling and inaccuracies in the measurements of $\dot{V}O_2$ or W. The work of breathing as usually calculated by the Campbell diagram does not take into account either the work done on the abdomen or the work of deformation on the rib cage[3]. Moreover, it has been suggested that during loaded breathing some postural muscles may become active, consuming an extra amount of oxygen. To test this hypothesis we measured the EMG activity, via surface electrodes of the following muscles: trapetius (T), latissimus dorsi (L), pectoralis (P), deltoid (D), sternocleidomastoid (S), obliques externus (O) and rectus abdominis (R). In 4 normal subjects during CO_2-induced hyperventilation without any external impedance and with an inspiratory resistance, the inspiratory EMG activity of the investigated muscles, expressed by the average rectified integrated EMG as % of the EMG max multiplied by the duty cycle, was compared at isowork values in inspiratory resistance loaded and unloaded rebreathing runs. Variability between subjects in the pattern of recruitment was observed. However, we found that some relatively large muscles of the trunk, such as the trapetius and the latissimus dorsi, had a consistent increase in EMG activity (at isowork) during the resistive breathing runs in all subjects (Fig. 6). The increased EMG activity indirectly reflects the increased oxygen requirement[6], confirming that efficiency may be reduced because of the increased recruitment of the postural muscles, which increase oxygen consumption but do not contribute to the development of inspiratory work.

References

1. AGOSTONI, E., CAMPBELL E.J.M., FREEDMAN S.. Energetics. In: Campbell E.J.M. (Ed.) The respiratory Muscles: Mechanics and Neural Control, Philadelphia, PA: Saunders, 1970, 115-142.
2. AGOSTONI, E., FENN W.O.. Velocity of muscle shortening as a limiting factor in respiratory airflow. J. Appl. Physiol. 1960. 15: 349-353.
3. AGOSTONI, E., MOGNONI G., TORRI G., MISEROCCHI G.. Forces deforming the rib cage. Respir. Physiol. 1966. 2: 105-117.
4. BARTLETT, R.G., Jr., BRUBACH H.F., SPECHT H.. Oxygen cost of breathing. J. Appl. Physiol. 1958. 12: 413-424.
35. BELLEMARE, F., GRASSINO A.. Effect of pressure and timing of contraction on human diaphragm fatigue. J. Appl. Physiol. 1982. 53: 1190-1195.
6. BIGLAND-RITCHIE, B., WOODS J.J.. Integrated EMG and oxygen uptake during dynamic contractions of human muscles. J. Appl. Physiol. 1974. 36: 475-479.
7. BRADLEY, M.E., LEITH D.E. Ventilatory muscle training and the oxygen cost of sustained hyperpnea. J. Appl. Physiol. 1978. 45: 885-892.
8. CAMPBELL, E.J.M., WESTLAKE E.K., CHERNIACK R.M.. Simple method of estimating oxygen consumption and efficiency of the muscles of breathing. J. Appl. Physiol. 1957. 11: 303-308.

9. CAMPBELL, E.J.M., WESTLAKE E.K., CHERNIACK R.M. The oxygen consumption and efficiency of the respiratory muscles of young male subjects. Clin. Sci. 1959. 18: 55-64.
10. CHERNIACK, R.M. The oxygen consumption and efficiency of the respiratory muscles in health and emphysema. J. Clin. Invest. 1959. 38: 494-499.
11. COURNAUD A., RICHARDS D.W., BAKER R.A., BAKER M.E., FISHMAN A.P.. The oxygen cost of breathing. Trans. Assoc. Am. Physicians. 1954. 67: 162-173.
12. FENN, W.O. A quantitative comparison between the energy liberated and the work performed by isolated sartorius of the frog. J. Physiol. (London) . 1923. 58: 175-203.
13. FIELD S., SANCI S., GRASSINO A. Respiratory muscle oxygen consumption estimated by the pressure-time index. J. Appl. Physiol. 1984. 57: 44-51.
14. FRITTS H.N., Jr. FILLER J., FISHMAN A.P., COURNAND A. The efficiency of ventilation during voluntary hyperpnea. J. Clin. Invest. 1959. 38: 1139-1348.
15. GOLDMAN M.D., GRASSINO A., MEAD J., SEARS T.A. Mechanics of the human diaphragm during voluntary contraction: Dynamics. J. Appl. Physiol. 1978. 44: 840-848.
16. GRASSINO, A., GOLDMAN M.D., MEAD J., SEARS T.A.. Mechanics of the human diaphragm during voluntary contraction: Statics. J. Appl. Physiol. 1978. 44: 829-839.
17. HILL., A.V. The effect of load on the heat of shortening muscle. Proc. R. Soc. Biol. 1964. 159: 297-318.
18. KONNO K., MEAD J. Static volume-pressure characteristics of the rib cage and abdomen. J. Appl. Physiol. 1968. 24: 544-548.
19. McGREGOR M., BECKLAKE M. The relationship of oxygen cost of breathing to respiratory mechanical work and respiratory force. J. Clin. Invest. 1961. 40: 971-980.
20. McKERROW, C.B., OTIS A.B. Oxygen cost of hyperventilation. J. Appl. Physiol. 1956. 3: 365-373.
21. MILIC-EMILI, J., PETIT J.M. Mechanical efficiency of breathing. J. Appl. Physiol. 1960. 15: 359-362.
22. MOMMAERTS, W.S. Energetics of muscular contraction. Physiol. Rev. 1969. 49 (3): 427-508.
23. REID, M.B., JOHNSON R.L., Jr. Efficiency, maximal blood flow and aerobic work capacity of canine diaphragm. J. Appl. Physiol. 1983. 54: 763-772.
24. ROCHESTER, D.F., BETTINI G. Diaphragmatic blood flow and energy expenditure in the dog. Effects of inspiratory airflow resistance and hypercapnia. J. Clin. Invest. 1976. 57: 661-672.
25. ROUSSOS, C.S., MACKLEM P.T. The respiratory muscles. N. Engl. J. Med. 1982. 307: 786-797.
26. ROUSSOS, C.S. Energetics. In: Roussos C., Macklem P.T. (Eds.) Lung Biology in Health and Disease. The Thorax. New York: Dekker, 1985, 437-492.

A. GRASSINO - C. FRACCHIA - C. RAMPULLA - L. ZOCCHI
Respiratory Muscles in COPD
Bi & Gi Publishers, Verona - Springer-Verlag, London, 1988

EMG evaluation of respiratory muscles

A. ARRIGO[1], R. CASALE[2], M. BUONCUORE[2]

[1] *Chair of Clinical Neurophysiology, University of Pavia, "C. Mondino" Foundation, Pavia, Italy*
[2] *Clinical Neurophysiology Department, Medical Rehabilitation Centre, "Clinica del Lavoro" Foundation, Montescano, Pavia, Italy*

SUMMARY: Neurophysiological evaluation of respiratory muscles is applied to traditional diagnosis, kinesiology and the study of localized respiratory muscle fatigue.
Examination of relevant parameters is based on electromyogram reading of surface or intraesophageal electrode recordings; kinesiological and local muscular fatigue evaluation obviously necessitate the application of appropriate statistical criteria and comparisons.
EMG signals must be isolated from cardiac impulse or separated from it by subsequent calculation, after which they are quantified by rectification or by determination of the Root Mean Square.
Power spectrum analysis is, however, applied with difficulty to respiratory muscle activity, complicated as it is by neighbouring synergic or platysma activity.
Such interference, while not prejudicing routine clinical evaluations, is detrimental to finer experimental requirements.

KEY WORDS: Respiratory muscle; respiratory muscle fatigue localized; Electromyogram; EMG signals; respiratory muscle activity; diaphragm myographic activity.

The functional evaluation of the respiratory muscles is not part of the routine work of clinical neurophysiology departments, coming more within the scope of respiratory physiology. Complex research programmes based on electromyographic techniques have, nevertheless, been carried out for some time in foreign neurophysiology departments. Within Italy, the Centre at Montescano has been to the fore in this respect.

Available procedures of neurophysiological examination can give information on respiratory muscle function, not only in conditions of normal activity but also when affected by a variety of diseases or by fatigue.

Stated simply, these techniques find three areas of application: traditional diagnosis, kinesiology and the study of localized respiratory muscle fatigue. The traditional diagnostic procedure, as for the skeletal muscles, comprises two stages: analysis of respiratory parameters and of the voluntary recruitment of the motor units during respiration; and the study of the conduction of the relevant nervous impulse along the phrenic nerve. The phrenic nerve can be studied by stimulation to the neck, on the posterior edge of the sternocleidomastoid muscle.

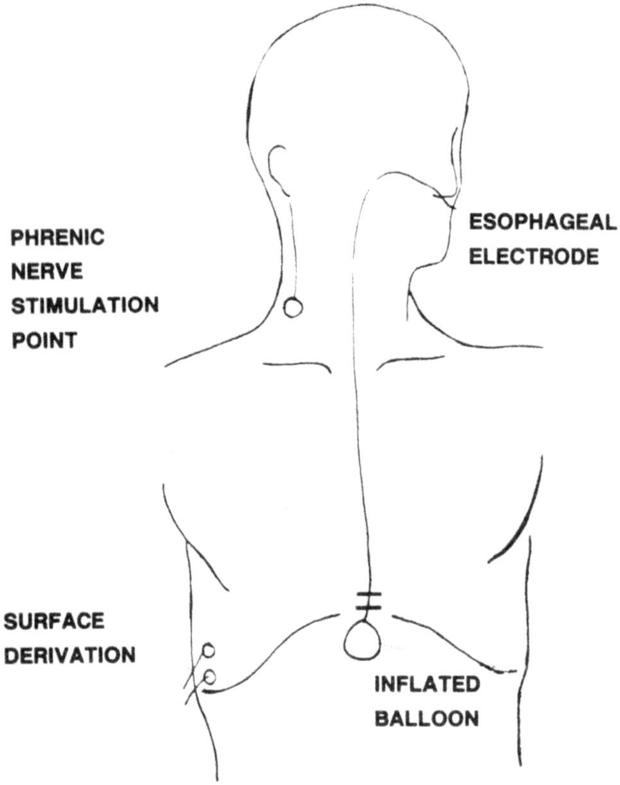

Fig. 1. - Stimulation point of the right phrenic nerve and methods of recording the diaphragm myographic activity[4].

Muscular response can be recorded with Ag/AgC1 surface electrodes placed in the 6th, 7th, 8th and 9th intercostal spaces, on the anterior axillary line. In this way it is possible to obtain the electrical activity of the homolateral hemidiaphragm costal fibres together with that of the intercostal muscles[8].

Response can also be measured by an intraesophageal electrode anchored and stabilised by an inflated balloon in the stomach (Fig. 1)[4]. This is the only electrode technique by means of which it is possible to read the electrical activity of the diaphragm pillars.

The muscular response to stimulation of the phrenic nerve can be classed as Mmax, on the basis of general criteria of stimulus and derivation[1]. By analyzing parameters of latency, amplitude and shape for this response, it is possible to identify conduction delays in the passage of the impulse along the phrenic nerve and/or any deficit of the muscular effector (diaphragm).

The study of the recruitment of the respiratory muscles in normal breathing can also be performed for the diaphragm by electrodes placed as described above

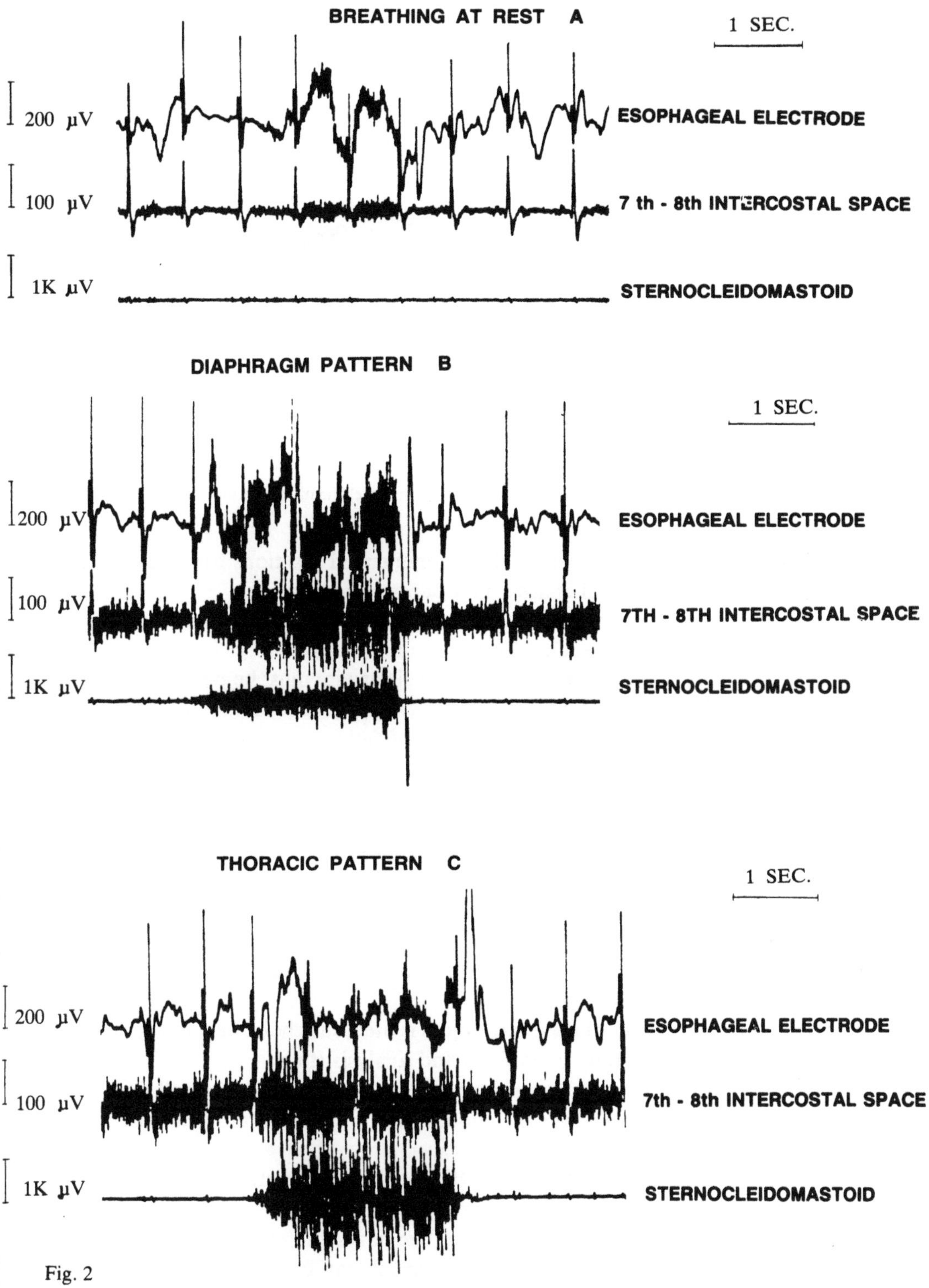
BREATHING AT REST A
1 SEC.
200 µV
ESOPHAGEAL ELECTRODE
100 µV
7 th - 8th INTERCOSTAL SPACE
1K µV
STERNOCLEIDOMASTOID
DIAPHRAGM PATTERN B
1 SEC.
200 µV
ESOPHAGEAL ELECTRODE
100 µV
7TH - 8TH INTERCOSTAL SPACE
1K µV
STERNOCLEIDOMASTOID
THORACIC PATTERN C
1 SEC.
200 µV
ESOPHAGEAL ELECTRODE
100 µV
7th - 8th INTERCOSTAL SPACE
1K µV
STERNOCLEIDOMASTOID

Fig. 2

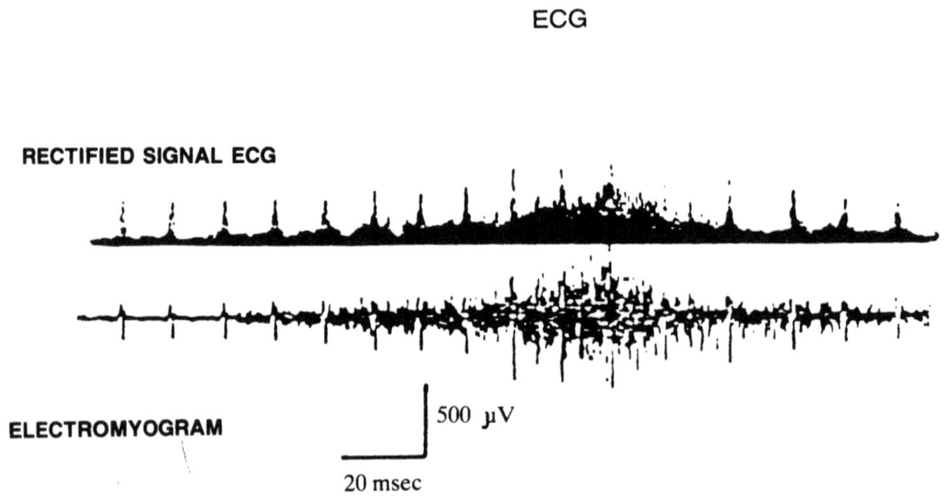

Fig. 3. - Model of EMG signal rectification as recorded from the left hemidiaphragm of a normal subject.

(Fig. 2A-B-C). For the intercostals, the recommended positions for study are the 2nd, 3rd, and 4th intercostal spaces, in a parasternal position.

It is hardly necessary to stress that the use of needle electrodes can create a risk of pleural puncturing as a result of brusque movements.

For all accessory muscles, surface or needle electrodes must be positioned on the muscle venter. Traditional techniques of electromyogram reading are used[2, 3].

For kinesiological evaluation and the study of fatigue, methods must be chosen with a view to allowing a statistical comparison of quantified parameters. In this respect, mere description of the EMG signal is insufficient. It is therefore necessary to use techniques suitable for the statistical evaluation and comparison of parameters relating both to kinesiology and to localized muscular fatigue.

A problem in quantitative EMG evaluation of the respiratory muscles is the constant presence of the cardiac impulse. There are two ways to eliminate this unwanted 'interference': either by reading the EMG signal alone between two cardiac signals, or by automatic computerized subtraction of the ECG signal from the reading, a process described by many authors.

In our laboratory, either of these techniques is used, the choice depending on their suitability to particular study conditions and analysis times.

Bearing in mind the general considerations so far presented, we will now pass to a more detailed analysis of the techniques used in our laboratory for the quantification of EMG parameters.

One technique is rectification of the EMG signal. Any myographic activity deflections, either above or below the isoelectric, are all carried above the relevant line (Fig. 3). The signal rectified in this way can then be integrated, after which the area circumscribed by the linear envelope of the graph can be calculated.

It can be useful to take into consideration only the linear envelope, in order to establish time ratio (kinesiologic parameters) between the activation of different muscles or successive activations of the same muscles, together with other, associated physiorespiratory parameters.

Another system of evaluation is to calculate the RMS (Root Mean Square), the mathematical expression of the square root of the average of the squares of a series of related values, both positive and negative, in a given recording period:

$$RMS = \frac{1}{N} \sum_{n=1}^{N} V_n^2$$

Since the EMG signal consists of a series of such values, the study of the RMS provides a useful parameter for the quantification of muscular electrical activity.

The method used by us to evaluate fatigue calls for the quantitative analysis of the frequency components of the signal. The theoretical basis to this analysis is that any analog signal can be broken down into a number of sinusoid components whose frequency is a multiple of basic frequency known as the fundamental harmonic of the signal. This decomposition is obtained by the application of an algorithm known as the Fourier transform. It allows the frequency of the initial analog signal, transformed to a digital scale with time domain passing to frequency domain, to be expressed in its various phases as a power spectrum. By plotting the mean values of the power spectrum at different times, it is possible to calculate the mean frequencies and/or centroids, on the basis of which variations of the frequency content of the signal under examination can be statistically compared[7, 8, 9, 10, 11, 12].

Grassino and other authors[8] prefer to take as an index of fatigue the high/low frequency content ratio of the spectrum. Localised muscular fatigue[5] can be expressed as the shift of the spectrum's contents to the lower frequencies, indicating that the particular effort under study can no longer be maintained.

Effort, measured for the skeletal muscles in kg by means of linear force transducers, is calculated with a T-valve nozzle whose resistance to forced inspiration can be varied.

EMG recording and its transformation into a power spectrum of the final signal is applicable only with difficulty to the study of respiratory muscles, because of the presence of synergic activity components produced by neighbouring muscles. An important example in this respect is the joint presence of diaphragm and intercostal muscle activity in an EMG signal obtained by surface derivations, with the added complication of platysma activity if an upper costal position is chosen for the electrodes.

It is also important to note that, if recording conditions are not varied (type of electrode used, their position, type and permeability of nozzle, type of inspiration or expiration), the inevitable complication of the signal with impulses from synergic muscles does not invalidate the interpretation of overall variations in the signal as parameters of fatigue. This is, in practice, the approach of clinical practitioners. For the finer requirements of certain laboratory measurements, the various signal components provided by the neighbouring muscles can be measured or calculated and subsequently removed.

References

1. ALFONSI E., MOGLIA A., SANDRINI G., PISONI M.R., ARRIGO A.: Electrophysiological study of long thoracic nerve conduction in normal subjects. Electromyogr. Clin. Neurophysiol 1986. 26: 63-67.
2. ARRIGO A.: Letture di Elettromiografia Clinica. Ed. EMI Pavia, 1982.
3. ARRIGO A.: Valutazione della fatica muscolare localizzata. Parte prima e parte seconda. Riv. Soc. It. EEG e Neurofisiol. Clin. 1 e 2, 1986.
4. BELLEMARE F., GRASSINO A.: Evaluation of human diaphragm fatigue. J. Appl. Physiol. Environ, Exercise Physiol. 1982. 53: 1196-1206.
5. CHAFFIN D.B.: Localized muscle fatigue-definition and measurement. J. Occup. Med. 1973, 15: 346-352.
6. DE LUCA C.J.: Myoelectric manifestation of localized muscular fatigue in humans. CRC Critical Review in Biomedical Engineering, 1985 II, 4: 251-279.
7. DE LUCA C.J., SABBHAI M.A., STULEN F.B., BILOTTO G.: Some properties of the median frequency of the myoelectrical signal 3 during localized muscle fatigue. In: Knuttgen H.K., Vogel J.A. Poortmans J. (Eds) Biochemistry of exercise. 1983.
8. GROSS D., GRASSINO A., ROSS W.R.D., MACKLEM P.T.: Electromyogram pattern of diaphagmatic fatigue, J. Appl. Physiol. Respirat. Environ. Exercise Physiol. 1979 46, 1-7.
9. LINDSTROM L., KADEFORS R., PETERSEN I.: An electromyographic index for localized muscle fatigue, J. Appl. Physiol. Respirat. Environ. Exercise Physiol. 1977, 43: 750-754.
10. MERLETTI R., SABBHAI M.A., DE LUCA C.J.: Median frequency of the myoelectric signal: effect of ischemia and cooling. Env. J. Appl. Physiol. 1983, 52: 258-265.
11. MILLS K.R.: Power spectral analysis of electromyogram and compound muscle action potential during muscle fatigue and recovery, J. Physiol. London 1982, 326: 401-409.
12. SCHWEITZER T.W., FITZGERALD J.S., BOWDEN A., LYNNE-DAVIES P.: Spectral analysis of human inspiratory diaphragmatic electromyograms. J. Appl. Physiol. Respirat. Environ. Exercise Physiol. 1979. 46, 152-165.

Aspects of Respiratory Muscles Pathophysiology

A. GRASSINO - C. FRACCHIA - C. RAMPULLA - L. ZOCCHI
Respiratory Muscles in COPD
Bi & Gi Publishers, Verona - Springer-Verlag, London, 1988

Pathways leading to skeletal muscle fatigue

A. GRASSINO
Meakins-Christie Laboratories, McGill University, and Notre-Dame Hospital, University of Quebec, Montreal, Canada

SUMMARY: Factors leading to development of respiratory muscle fatigue include:
- high inspiratory loads;
- weak inspiratory muscles: neural disease, partial paralysis, disuse atrophy; malnutrition; high CO_2, low O_2, hypophosphatemia, chronic fatigue;
- low blood perfusion to the respiratory muscles;
- pattern of breathing consisting of long T_I/T_T, high inspiratory flow (increased inspiratory resistance, exercise, respectively)
- high tidal volume;
- defective coupling between rib cage and diaphragm.

KEY WORDS: Skeletal muscle fatigue; respiratory muscle fatigue; high inspiratory loads; weak inspiratory muscle; neural disease; partial paralysis; disuse atrophy; malnutrition; high CO_2; hypophosphatemia; chronic fatigue; low blood perfusion; breathing; high inspiratory flow; inspiratory resistance exercise; tidal volume; defective coupling; rib cage; diaphragm; overinflated lungs; nutritional factors.

An illustrative definition of skeletal muscle fatigue and its development is represented by the familiar sequence of events triggered when a heavy suit case is lifted. Lifting the suit case imposes a load to the arm and shoulder muscles; the subject experiences a progressive sensation of numbness, then pain in the arm, the shoulder muscles are recruited to help lifting, the suit case feels heavier and finally falls to the floor. This failure of the arm muscles to keep the suit case off the ground is called fatigue, implying mechanical failure. Usually the suit case is picked up by the other arm and the walk continues. Why was the suit case dropped?

Several determinant factors in the development of skeletal muscle fatigue and failure can be identified by answering some basic questions:

1) Where did failure occur?

In such a voluntary effort it is not certain whether the force that can be exerted is limited by the capacity of the nervous system and/or conducting pathways to deliver motor impulses to the muscle (central fatigue) or by alteration of the excitation-contraction coupling of the muscle fibers (peripheral fatigue). It thus remains to be ascertained whether a voluntary effort can be improved by supramaximal tetanic electrical stimulation of the muscle. Often in fatigue it is not clear if the tension falls because the degree of voluntary innervation drops or because the fibers are incapable of maintaining their contraction. Merton[1] was among the first to find that fatigue is peripheral, for when strength fails, electrical stimulation of the motor neurons cannot restore it. Merton further observed that recovery from fatigue does not take place if the circulation to the muscle is interrupted.

Recently, Bellemare and Bigland-Ritchie proposed that fatigue of the diaphragm may be an exception to the rule and could be induced by decrease in its central activation[2].

2) How heavy has the load to be?

Mixed fiber type skeletal muscles fail if the load imposed on them exceeds about 15 to 20% of their maximal force (in near isometric contractions), i.e. there is a relationship between maximal muscle force and the magnitude of the load that can be carried. Heavier loads result in earlier failure. Muscles of different fiber types have different tolerance to loads, the fast fiber type fatiguing early. The fatigue threshold will depend as well on the velocity of shortening of the muscle, with faster contraction resulting in earlier fatigue.

3) Why do muscles fail?

If the fatigue occurs at muscle level, an important determining factor is the availability of blood circulation to the muscle, both because it washes out catabolites generated by the fiber's contraction and because it provides nutrients and O_2 to the muscles.

Fatigue is accompanied by a fall in phosphoryl creatinine and a rise in muscle lactate concentration. However, ATP concentration remains high, even in extreme fatigue. Accumulation of muscle lactate leads to a fall in muscle pH from 7 to as low as 6.4. This pH slows down the rate of ATP resynthesis from anaerobic glycolysis by inhibition of enzymatic reactions. In vitro studies show that the enzymes responsible for glycolysis are almost completely inactive at pH of 6.4.

4) How long does fatigue last?

Fatigue is a reversible physiological situation and force recovers after a period of rest, when blood flow is reinstituted to the muscle.

While fatigue development in the respiratory muscles is more complex than the above example, the basic rules of energy demand and supply apply to them as to the other skeletal muscles[3, 4].

Fatigue is a complex series of changes in the muscle fiber physiology and each parameter measuring a function shows a particular time constant of recovery, as shown in Figure 1. Recovery of maximal force is progressive and is completed in 10-15 minutes. Instead, the capacity to reproduce the initial endurance time is recovered at a much slower rate, possibly only in 18-24 h. Changes in the EMG frequency spectrum are recovered within 2-3 minutes. The relaxation rate of the muscle also recovers in about 2-3 minutes. The force response to supramaximal stimulation at 20 and 100 Hz shows different rates; force at 100 Hz is recovered

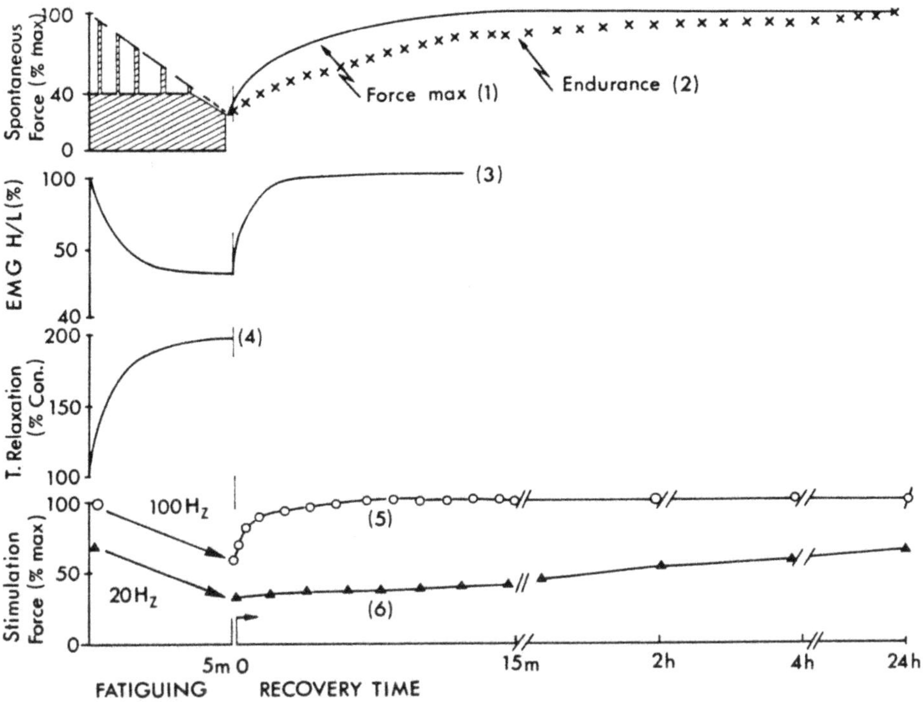

Fig. 1 - Schematic time course of various parameters during an isometric, sustained, fatiguing contraction and its recovery in a skeletal muscle. (Upper panel) Voluntarily sustained contraction at 40% of maximal force. Spikes show progressive loss of maximal force. T_{lim} occurs when Fmax = sustained force. Recovery of the capacity to develop maximal force (1) and to achieve T_{lim} (2) are shown.
(Second panel) Indicates the time course of the H/L of the EMG of the human diaphragm holding a contraction similar to that shown above (3). (Third panel) Shows the time course of the relaxation time (4) of the human diaphragm. Control value is in fresh muscle.
(Bottom panel) Shows decay of the force obtained by brief supramaximal stimulation with 20 and 100 Hz during a fatiguing contraction and recovery. Notice that, in fresh muscle, force developed with a 20 Hz stimulation is considerably smaller than that with 100 Hz. The rate of recovery of the latter is longer (5, 6).

in about 5 m, while force elicited by 20 Hz may take several hours to recover (low frequency fatigue).

Determinants of inspiratory muscle fatigue

1) Respiratory muscles contract intermittently, i.e. inspiratory muscles are loaded during inspiration and mainly relaxed during expiration, allowing them a period of rest and recovery. The relevance of this factor is shown in Figure 2, depicting the various combinations of inspiratory time/total duration of the cycle and force developed by the diaphragm. The isopleth (dashed line) includes all breathing patterns with an isoendurance of about 1 h and shows all possible combinations of

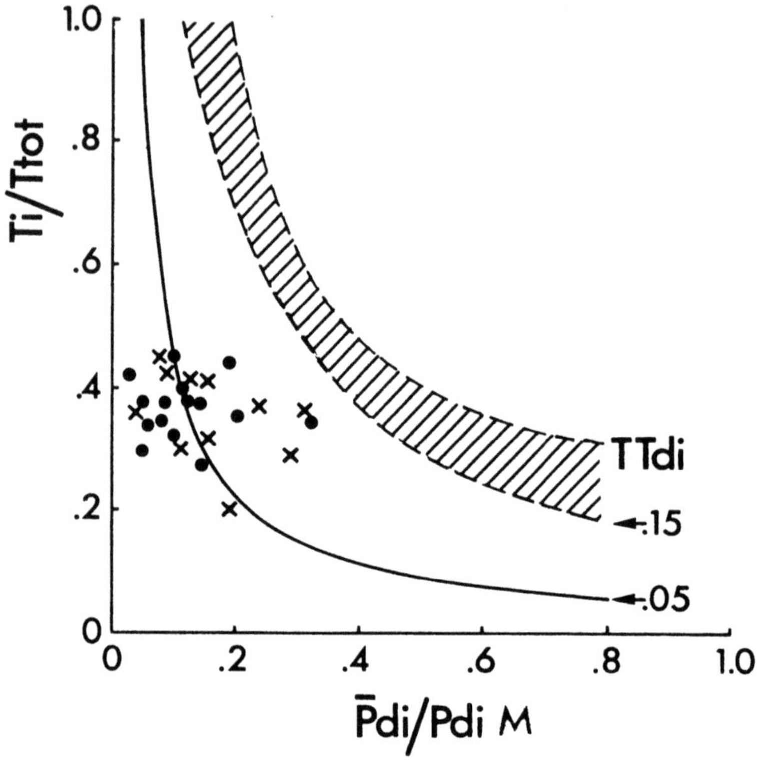

Fig. 2 - Relationship between inspiratory time cycle duration and mean Pdi of resting breaths/Pdi-max obtained in a group of COPD patients with mild to severe airways obstruction. The isopleth with dashed lines represent a TTdi of 0.15 - 0.18, the area of effort at which fatigue of the diaphragm starts to develop. All breathing patterns falling to the left of the dashed line are not expected to develop fatigue. Those to the right will develop fatigue. Each data point is the breathing pattern of one patient. Force reserve of some patients is shown to be very limited (close to the fatigue threshold).[17]

Pdi and T_I/T_T leading to it (TTdi = 0.15 - 0.18). High force can be sustained for 1 hour only if the relaxation time is large (short inspiration, long expiration)[5].

On the other hand, small efforts, about 15% of max, can be sustained for about one hour only, because the contraction is continuous ($T_I/T_{TOT} = 1$). Patterns of breathing falling on the left side of the shaded area do not lead to fatigue. Patterns falling to the right result in fatigue. Each data point in Figure 2 represents average resting breathing Pdi and T_I/T_{TOT} in one COPD patient[5].

2) Coupling of the respiratory muscles. The chest wall was described by Konno and Mead as having 2 degrees of freedom, i.e. we can inspire with the rib cage, leaving the diaphragm and abdomen more or less relaxed. On the other hand, we can sustain breathing by using only the diaphragm (as quadriplegic patients do), leaving rib cage muscles almost inactive. Often we use a combination of rib cage and diaphragm, and the proportion varies with body posture, constraints applied

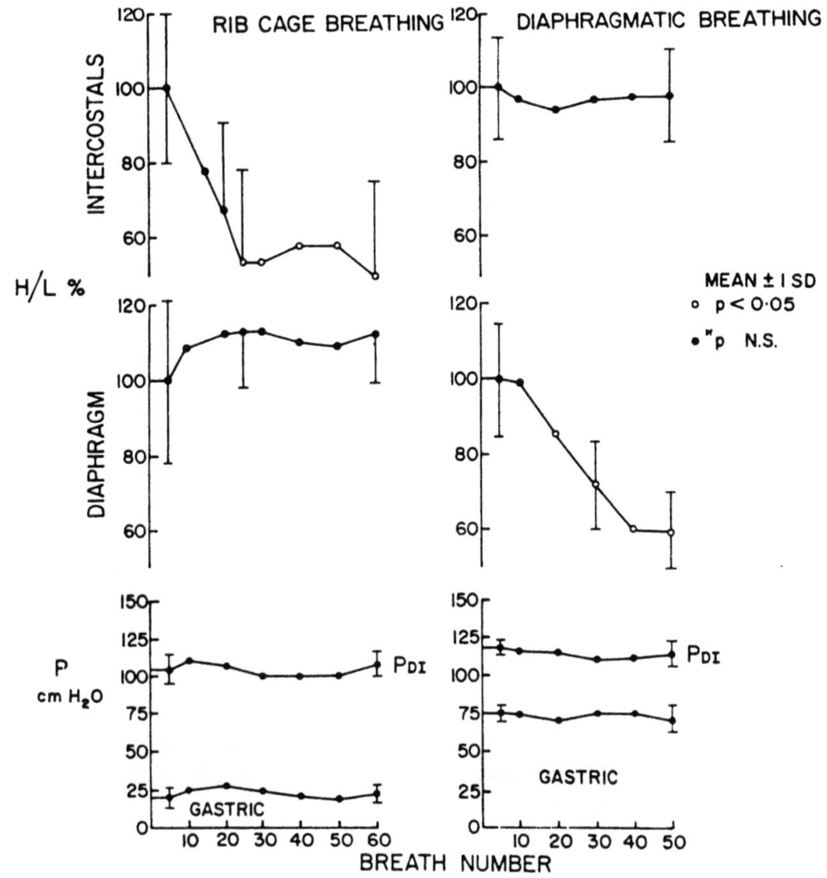

Fig. 3 - Relationship between H/L ratio of parasternal muscles and diaphragm EMG in normal subjects breathing at iso Pdi with predominant use of the diaphragm or predominant use of the rib cage-accessory muscles. Notice that selective fatigue of either muscle can be obtained by volitional coupling of these two muscle groups. [6]

to each, breathing stimuli, etc. Hence, a given inspiratory load can be sustained mainly by voluntary activity of rib cage muscles, or the diaphragm, or alternatively by both of them. Recent work by Fitting et al.[6] shows that fatigue of the rib cage or diaphragm can be elicited separately in normal subjects, depending on what group of muscles is selectively activated (Fig. 3).

If resistive breathing is held with rib cage muscles while Pg is low, we note they develop fatigue, while the diaphragm will not show changes. The same inspiratory resistance, if sustained with the diaphragm, will fatigue it, sparing the rib cage muscles. EMG frequency spectrum changes obtained from esophageal electrodes for the diaphragm and fine wire electrodes for the intercostals and sternocleidomastoid and pectoralis were used as an index of developing fatigue. The degree of contribution of the rib cage muscles or diaphragm can be assessed by the relative swing of Ppl and Pg. We do not yet know the fatigue threshold for rib cage muscles. When the diaphragm is loaded, both the costal and crural portions develop fatigue.

Velocity of shortening (inspiratory flow)

While there is uncertainty concerning the relationship between velocity of shortening of intercostal muscles and mouth flow (in dogs), the relationship with diaphragmatic shortening is much better established. Fast muscle contractions consume more oxygen than slow contractions and in addition reduce the functional maximal force of the muscle. As shown by Agostoni and Rahn, Pdimax at a flow of about 3 l/sec is about 25% smaller than that obtained statically. Hence, force reserve is down at high inspiratory flow and the fatigue threshold shown in Figure 2 (obtained at a flow below 1 l/sec) will probably be displaced to the left. There is experimental evidence indicating this statement is correct in normal subjects during exercise-induced hyperpnea. Hussain and Pardy[7] exercised normal subjects on a bicycle ergometer at 80% of their maximal power output to exhaustion, with and without rib cage strapping. EMG signals of fatigue developed at TTdi of about 0.09. Under such conditions, V_T/T_I was up to 4.5 l/sec, indicating a considerably high velocity of shortening for the inspiratory muscles. At such a flow, however, it is expected that Pdimax will be about 35% to 40% lower than that used by Hussain and Pardy to calculate TTdi. Hence, if the prevailing Pdimax at high flows were to be used as Pdimax (rather than static Pdimax), the TTdi in those experiments would again be in the neighbourhood of the values described by Bellemare and Grassino as the fatigue threshold[5, 17].

Tidal volume

Large tidal volumes result in an increase in inspiratory muscle shortening, also necessitating the development of higher muscular force to overcome the increased elastic recoil and resistive load of the lungs. There is a likelihood that fatigue will develop earlier, as demonstrated in obese patients during CO_2 rebreathing. However, careful studies analyzing those factors in humans have yet to be made available.

Overinflated lungs

COPD or emphysema usually flattens the diaphragm, as shown by X-Rays. As suggested by Macklem et al., this alters the action of the diaphragm on the rib cage from the inspiratory action of lifting the rib cage to the expiratory action of pulling inwards its lateral diameter. This is to say that, even if force can still be generated by the diaphragm, it is applied in the wrong direction to make it function as an inspiratory pump[8]. Arora and Rochester[9] recently assessed the effect of COPD on diaphragm muscle dimensions by measuring thickness, area and length in a group of COPD patients at necropsy, and compared them with data from sex-size matched non-COPD subjects. In COPD patients (LTC = 135% predicted) diaphragm mass was 213 G, thickness 3.2 mm and area 647 sq. cm, values not significantly different from those obtained in non COPD patients. No correlation was found between diaphragm length and lung volume in 70% of the patients, an indication that in COPD there is no evidence for the permanent shortening of the diaphragm previously suggested in a hamster experimental model[10, 11]. It was also confirmed that diaphragm muscle mass and body weight are linearly related both in male and female.

Nutritional factors

Dr. Rochester and co-workers found that malnutrition, expressed as loss of body weight, reduces respiratory muscle weight and force[12]. Similarly, lack of activity of respiratory muscles given by lack of use (prolonged mechanical ventilation) will decrease their mass and force. Nutrition and exercise are dynamic variables which can be influenced by treatment. Hypophosphatemia and high CO_2 also reduce muscle performance.

Fatigue

The development of fatigue reduces the maximal force a muscle can generate and impairs its contractility. A muscle forced to contract chronically against heavy loads may develop a state of chronic fatigue (and weakness). Recovery from such a state is slow and takes several hours to days, as shown in Figure 1[3].

Blood perfusion to respiratory muscles

Blood flow to muscles can be limited by lower perfusion pressure[13] (cardiac tamponade, hemorrhagic shock) or by strong muscle contraction, closing the muscles' capillary bed. Vascular resistance is another regulatory factor. The normal increase in diaphragm blood flow induced by loads seems to occur only if perfusion pressure is above 60 mmHg[15].

The circulation of the diaphragm is rather special. Comptois et al.[16] described it as formed by the anastomosis of arteries branching from the subclavian artery (via the internal mammary) to several intercostals and the phrenic artery branching from the abdominal aorta. It is of interest that these arterial anastomoses form a ring around the central tendon. From this ring, there are emerging branches that form anastomoses connecting with the intercostals (Fig. 4).

Bellemare et al.[13] canalized a phrenic vein and measured its blood under various patterns of contraction of the diaphragm. They found that low force contractions increase blood flow, while strong contractions will result in limitation of blood flow during contraction (inspiration) and hyperemia during relaxation (expiration). In dogs with casted abdomen it was further found that there is a unique relationship

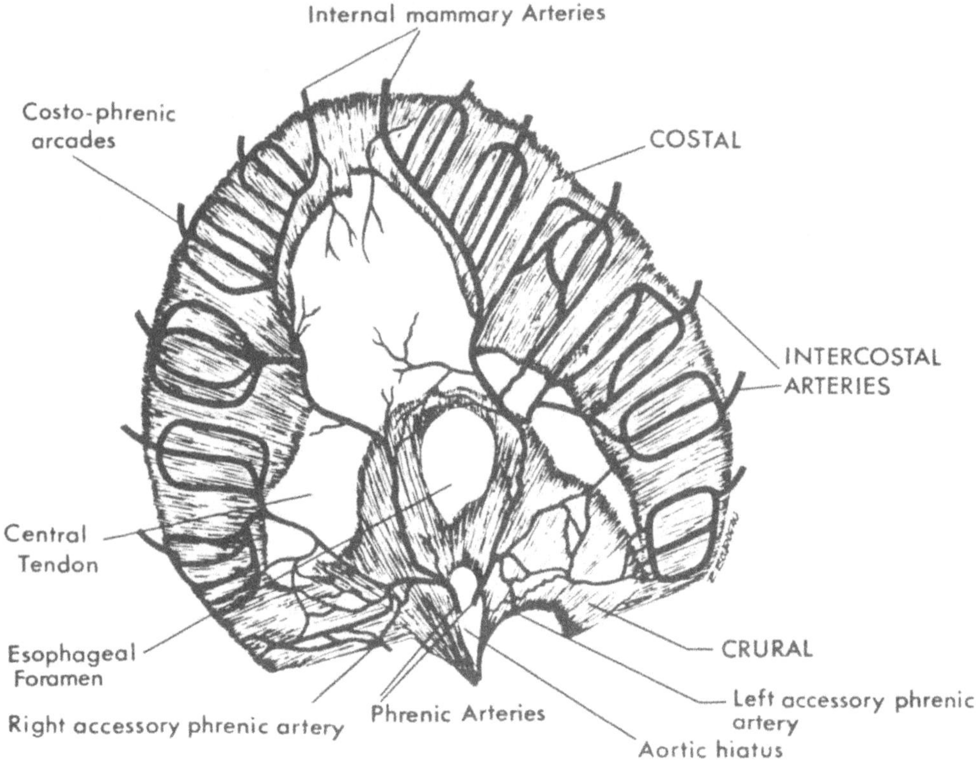

Fig. 4. - Abdominal view of diaphragm illustrating the internal circle formed by the phrenic and internal mammary arteries. In addition, branches of intercostal arteries communicating via costophrenic arcades with the internal circle. Large dark arrows show the three main arterial supplies to the diaphragm. [16]

between the tension time (TTdi) of the diaphragm and the level of blood flow. At TTdi about 0.15, post-contraction hyperemia starts to be seen. Blood flow increases in absolute values up to a TTdi of 0.25, then decreases, while post-exercise hyperemia consistently increases. There is blood flow interruption when Pdi reaches about 70% of maximal (Fig. 5).

Work by Butchler et al.[14] shows that blood flow increases with frequency of breathing, and may vary if the transdiaphragm pressure is generated mainly by increases in Ppl or Pg. These are interesting factors pointing to a complex hemodynamic equilibrium in the diaphragm.

Who are the patients at risk of developing fatigue of respiratory muscles?

Schematically, we can identify three groups: 1) those who have low respiratory muscle force, whatever the cause; 2) those who have low lung or chest wall compliance; and 3) the most common group, a mixture of both[3].

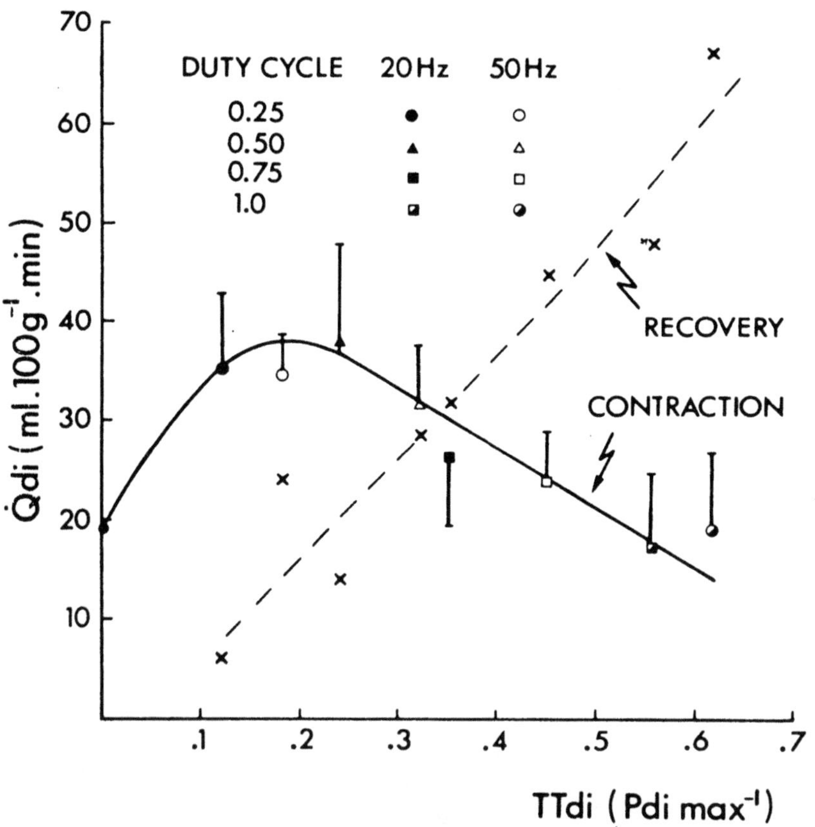

Fig. 5. - Relationship between final diaphragmatic blood flow (Qdif) during a contraction period, debt measured during recovery, and diaphragmatic tension-time index (TTdi) in 4 animals during periodic contractions with duty cycles of 0.25, 0.75, and 1.0. Results obtained at 20 and 50 Hz are shown. Bars indicate SE for contraction; during recovery SE is 5.85 at 20 and 17.4 at 50 Hz. Best-fitted lines by eye are drawn. For further explanation see text. [14]

Bellemare et al. show in a group of COPD patients that their TTdi range from near normal to very close to the fatigue threshold, and demonstrate that changes in breathing timing or force could bring a borderline patient into fatigue[17].

Begin et al.[18] reviewed 242 COPD patients who underwent respiratory mechanic tests over a 3-year period and drew the following conclusions (Fig. 6):

RL is an index of the load offered by the lung to the muscles, and is an expression of lung damage.

MIP is an index of inspiratory muscle force reserve. The combination of high resistance and low MIP seems to define the group of patients at higher risk, while low RL and high MIP are typical of normal subjects.

The ratio RL/MIP is an expression of the degree of load/force reserve. The higher the numerical values, the lower the reserve.

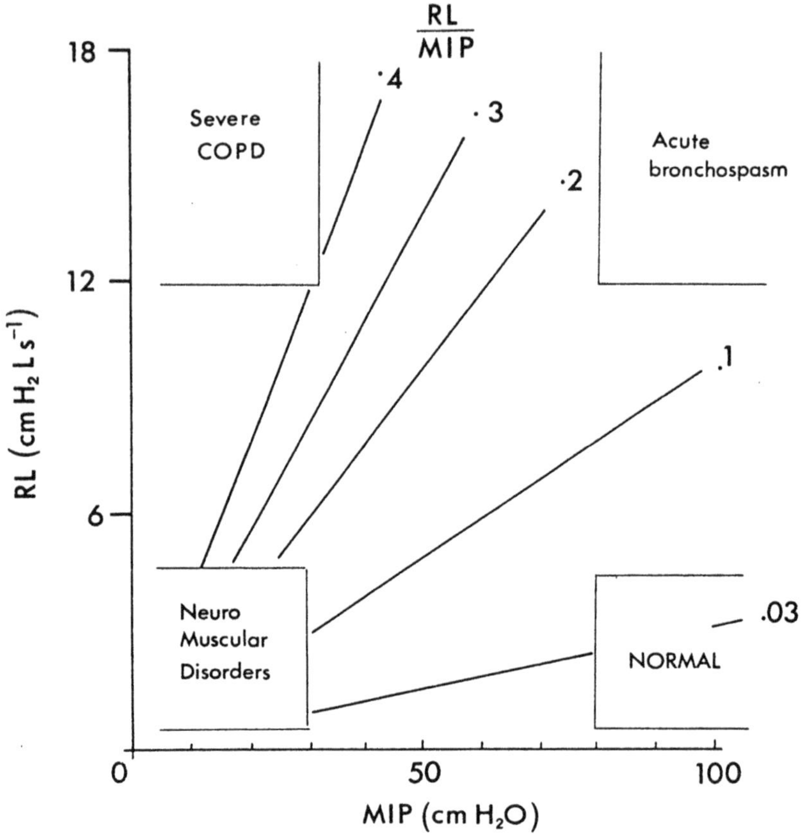

Fig. 6. **Relationship between pulmonary resistance measured during resting breathing and maximal inspiratory pressure at FRC. Isopleths (RL/MIP) represent a relationship between force required to breathe and maximal force available. The four square areas indicate where specific diseases may be placed. In a population of 242 COPD patients, all those with an index of 0.35 or higher had CO_2 retention.** [18]

When the RL and MIP values of the 242 COPD patients are plotted, there is a great dispersion of values depending on multiple factors. However, those who retain CO_2 chronically ($PaCO_2$ above 45 mmHg), while present all over the RL/MIP diagram, constitute a strong majority of patients who have an index of about 0.4; this suggests that the combination of low inspiratory muscle force and high lung resistance leads invariably to CO_2 retention and makes patients combining these factors the most likely to develop fatigue and acute muscular failure.

References

1. MERTON P.A. Voluntary strength and fatigue. J. Physiol. 1954. 123: 553-64.
2. BIGLAND-RITCHIE B., BELLEMARE F., WOODS JJ. Excitation frequencies and site of fatigue in human muscle power: factors underlying maximal performance. Human Kinetics Publishers 1985.
3. GRASSINO A., MACKLEM P.T., Respiratory muscle fatigue and ventilatory failure. Ann. Rev. Med. 1984. 35: 625-47.
4. ROUSSOS C., MACKLEM P.T., Inspiratory muscle fatigue. In: Handbook of Physiology, Section 3. The Respiratory System. Vol. III, part 2, 1986; 511-527.
5. BELLEMARE F., GRASSINO A., Evaluation of human diaphragmatic fatigue. J. Appl. Physiol. 1982. 53: 1196-1206.
6. FITTING J.W., BRADLEY T.D., EASTON P.A., GOLDMAN M.D., GRASSINO A. Dissociation between diaphragmatic and rib cage muscle fatigue. Am. Rev. Respir. Dis. 1986. 1986, A133, 4, A251.
7. HUSSAIN S., PARDY R.L. Inspiratory muscle function with restrictive chest wall loading during exercise in normal humans. J. Appl. Physiol. 1985. 58: 2027-32.
8. MACKLEM P.T., MACKLEM D.M., DE TROYER A. Model of inspiratory muscle mechanics. J. Appl. Physiol. 1983. 55: 547-57.
9. ARORA N.S., ROCHESTER D.F. Effect of chronic obstructive pulmonary disease on human diaphragm muscle dimension. Chest 1986; In Press.
10. FARKAS G.A., ROUSSOS C. Adaptability of the hamster diaphragm to exercise and/or emphysema. J. Appl. Physiol. 1982. 53: 1263-1272.
11. SUPINSKY G. KELSEN S. Effects of elastase-induced emphysema on the force generating ability of the diaphragm. J. Clin. Invest. 1982. 70: 978-88.
12. ROCHESTER D.F., BRAUN N.M., ARORA N.S. Respiratory muscle stregth in COPD. Am. Rev. Respir. Dis. 1979. 119: 151-54.
13. VIIRES N., SILLY M., AUBIER M., RASSIDAKIS A., ROUSSOS C. Effects of mechanical ventilation on respiratory muscles and regional blood flow distribution during cardiogenic shock. J. Clin. Invest. 1983. 72: 935-47.
14. BELLEMARE F. WIGHT D., LAVIGNE C., GRASSINO A. Effect of tension and timing of contraction on the blood flow of the diaphragm. J. Appl. Physiol. 1983. 54: 1597-1606.
15. BUCHLER B., MAGDER S., ROUSSOS C. Effects of contraction frequency and duty cycle on diaphragmatic blood flow. J. Appl. Physiol. 1985. 58: 265-73.
16. COMPTOIS A., GORCZYCA W., GRASSINO A. Anatomy of the diaphragmatic circulation. J. Appl. Physiol. In Press, 1986.
17. BELLEMARE F., GRASSINO A. Force reserve of the diaphragm in COPD. J. Appl. Physiol 1983. 55: 8-15.
18. BEGIN P., GRASSINO A. Prevalence of CO_2 retention in COPD patients related to inspiratory muscles and lung dysfunction. Am. Rev. Respir. Dis. 1986; A133, 4, A191.

A. GRASSINO - C. FRACCHIA - C. RAMPULLA - L. ZOCCHI
Respiratory Muscles in COPD
Bi & Gi Publishers, Verona - Springer-Verlag, London, 1988

Efficientcy of breathing during hyperinflation

P.W. COLLET[1], L.A. ENGEL[2]

1. *Laboratory of Experimental Medicine, Faculty of Medicine, University of Aix-Marseille, France*
2. *Thoracil Medicine Unit, Westmead Hospital, Sydney, Australia*

SUMMARY: The adverse effects of hyperinflation on respiratory muscle function are:
- increased work of breathing;
- decreased inspiratory muscle strength;
- inspiratory muscle blood flow impairment;
- greater proneness of the inspiratory muscles to fatigue.

The O_2 cost of inspiratory resistive breathing in 30 paired runs performed by 5 trained normal subjects proved greater at high V_L than at FRC, while higher efficiency was obtained at FRC.
Reduced efficiency of breathing in hyperinflation may be related to:
- greater postural or stabilizing muscle recruitment;
- changes in inspiratory muscle mechanical coupling;
- decreased intrinsic inspiratory muscle efficiency.

Hyperinflated patients may thus, in the presence of such factors as airflow obstruction, suffer considerably reduced inspiratory muscle endurance.

KEY WORDS: Breathing; hyperinflation adverse effect; respiratory muscle function; inspiratory muscle mechanical coupling; inspiratory muscle efficiency; inspiratory muscle endurance.

Acute hyperinflation is disadvantageous for the respiratory muscles. First, the oxygen cost of breathing ($\dot{V}O_2$ resp) is increased. This is because the work of breathing is increased, reflecting increased elastic work at higher lung volumes, as well as increased resistive work if there is airway narrowing[11]. Second, inspiratory muscle strength decreases with increasing lung volume[4,5,9,13,15], because of the intrinsic length-tension relationship of the respiratory muscles[6,7,12] or changes in mechanical coupling at high lung volume[8]. Third, inspiratory muscle blood flow may be impaired both by greater intramuscular pressures during inspiration and by limitation of blood flow as a result of persistent muscle activity during expiration[10,14]. The latter is potentially more detrimental because inspiratory muscle blood flow is greatest during expiration[1]. Fourth, the inspiratory muscles are more prone to fatigue during hyperventilation[16]. This may reflect each of the first three disadvantages above. Thus, the need to generate a greater fraction of maximum available inspiratory pressure and blood flow limitation may both contribute to inspiratory muscle fatigue. An increase in the O_2 cost of breathing may also be critical, particularly if alveolar ventilation is limited by mechanical factors such as airflow obstruction.

During inspiratory resistive breathing in humans at FRC with constant tidal volume, the O_2 cost of breathing is mainly determined by the work rate[3]. This means that at constant tidal volume and frequency, the main determinant of the O_2 cost of breathing is the inspiratory pressure. Changes in respiratory timing or breathing frequency are only important to the extent that they affect work rate per minute. Moreover, with this type of breathing, efficiency is constant over a large range of loads, inspiratory flow rates and work rates. In contrast, in a recent study we have shown that the efficiency of breathing was reduced during acute hyperinflation[2].

The $\dot{O}_2$ cost of inspiratory resistive breathing at a given work rate was measured at two different lung volumes in 5 trained normal subjects. Paired loaded runs were performed at FRC and at an expiratory lung volume of FRC plus $37 \pm 2\%$ (mean SE) of inspiratory capacity (high V_L). For each pair of runs, tidal volume (0.6 liters), frequency (23 breaths/min), inspiratory flow rate (0.45 1/s) and inspiratory muscle pressure (45% of maximum static pressure at FRC) were kept constant, so that the pressure-time integral of inspiratory mouth pressure ($\int Pdt$) and the work rate across the external resistance were also constant. High V_L was maintained by continuous positive airway pressure (CPAP) of 9 ± 2 cm H_2O applied to a bag-in-box system by means of a clean air blower. If high V_L were maintained voluntarily the respiratory muscles would face an increase in internal elastic work. We eliminated this increase in internal elastic work by maintaining high lung volume passively, with externally applied pressure. The subjects were coached to relax during expiration to minimize respiratory muscle activity and allow the respiratory system to reach static equilibrium at end-expiration. The pressure in the bag-in-box system was adjusted so that the subject was breathing at the desired lung volume. Subsequent analysis showed that the values of end-expiratory pressure and lung volume fell on the respiratory system relaxation curve, indicating adequate relaxation.

At constant pressure-time product and work rate, the O_2 cost of breathing was greater at high V_L than at FRC in 28 of 30 paired comparisons. Thus, when mean lung volume (at mid-tidal volume) increased from $45 \pm 4\%$ of vital capacity (VC) for runs at FRC to $66 \pm 4\%$ VC for runs at high V_L the O_2 cost of comparable work performed by the inspiratory muscles increased from 109 ml O_2/min by $41 \pm 11\%$ ($p < 0.05$, Fig. 1).

Efficiency (work rate/$\dot{V}O_2$ resp) was 3.9 ± 0.2 % at high V_L compared to 5.2 ± 0.3 % at FRC ($p < 0.05$, Fig. 2). Thus an increase in mean lung volume equal to 21% of VC resulted in a 33% increase in the O_2 cost of breathing at a given work rate. When the work rate of breathing at high V_L was normalized for the decrease in maximum inspiratory muscle pressure with lung volume, efficiency at high V_L ($5.2 \pm 0.3\%$), calculated using the normalized work rate, did not differ from that at FRC ($p > 0.7$, Fig. 3).

Thus, the efficiency of breathing is less during acute hyperinflation. There are three possible mechanisms for this. First, a greater recruitment of muscles which

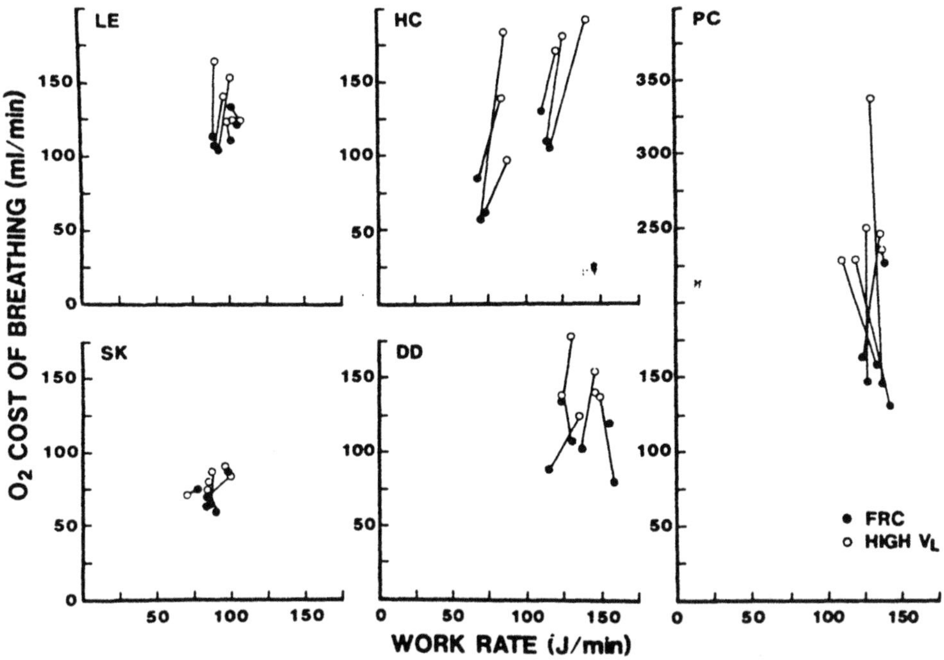

Fig. 1. - Relationship between O_2 cost of breathing ($\dot{V}O_2$ resp) and total work rate (work performed on external resistance, internal elastance, and thoracic gas decompression) in 5 subjects breathing with an inspiratory resistance at functional residual capacity (FRC) and at high lung volume (high V_L). Each point represents a single run; lines join paired runs. In 28 of 30 pairs $\dot{V}O_2$ resp was greater at high lung volume than at FRC. [2]

contract isometrically or have a postural, stabilizing or fixating function at high lung volume may account for the findings. If these muscles consume oxygen without shortening and performing work, overall efficiency would fall. Second, changes in the mechanical coupling of the inspiratory muscles during acute hyperinflation may be such that a given pleural pressure requires a greater fraction of maximum muscle tension. If O_2 cost depends on this fraction, the apparent efficiency would decrease. Increases in unmeasured work of breathing associated with changes in chest wall configuration would have a similar effect. Third, the findings may be explained by a decrease in the intrinsic efficiency of the inspiratory muscles when operating at higher lung volume and shorter length, analogous to the decrease in inspiratory muscle strength with increasing lung volume. The fact that efficiency at high V_L, when normalized for the fall in muscle strength with lung volume, was no different from that at FRC suggests that the fall in efficiency may be related to the fall in strength, though it does not allow one to differentiate between the different mechanisms. In the absence of detailed knowledge of the recruitment, force developed, mechanical coupling and shortening of the different muscles contributing to inspiration, one

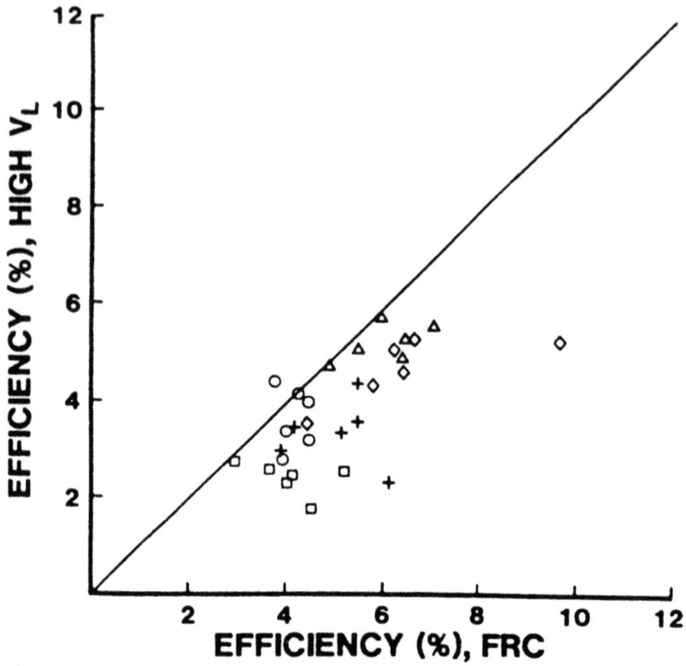

Fig. 2. - Identity plot of efficiency of breathing for paired runs at functional residual capacity (FRC) and high lung volume (high V_L). Symbols represent subjects. Efficiency was less at higher mean V_L. [2]

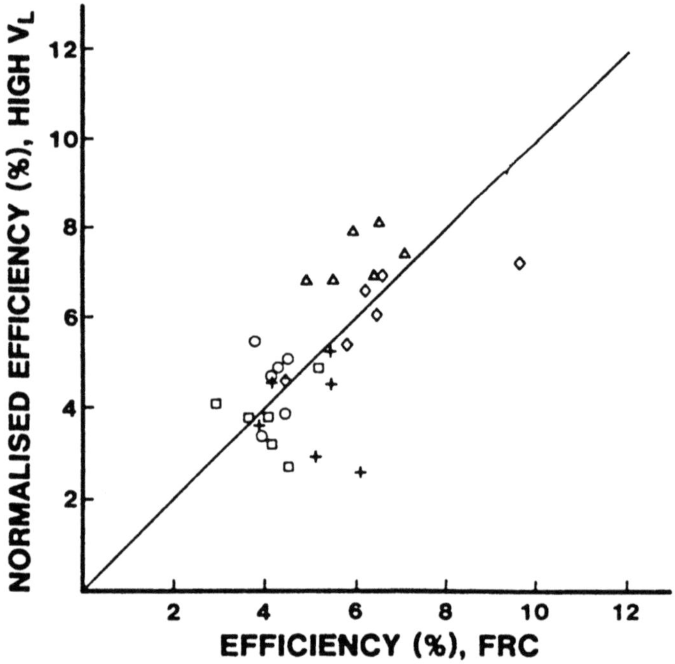

Fig. 3. - Identity plot for paired runs comparing efficiency at functional residual capacity (FRC) with normalized efficiency at high lung volume (high V_L). Normalized efficiency is efficiency multiplied by the ratio of maximum inspiratory muscle pressure (P max) at FRC to the value of P max at mean V_L during breathing at high V_L. Symbols represent subjects. [2]

can only speculate about the mechanisms for the decreased efficiency. It is possible that they all may contribute to the findings.

In conclusion, during acute hyperinflation, the O_2 cost of breathing is increased because of decreased efficiency. This means that the O_2 cost of breathing is further increased in clinical situations where it is already high because of increased work of breathing. In these circumstances, if the oxygen supply to the inspiratory muscles is critical, for example when ventilation and O_2 uptake by the lung is limited by factors such as airflow obstruction, a further increase of the O_2 cost of breathing may limit performance and contribute to the reduction in inspiratory muscle endurance[16].

Acknowledgement. This work was supported by the National Health and Medical Research Council of Australia.

References

1. BELLEMARE F., WIGHT D., LAVIGNE C.M., GRASSINO A. Effect of tension and timing of contraction on the blood flow of the diaphragm. J. Appl. Physiol. 1983. 54: 1597-1616.
2. COLLETT P.W., ENGEL L.A.. The influence of lung volume on the oxygen cost of resistive breathing. J. Appl. Physiol. 1986. 61: 16-24.
3. COLLETT P.W., PERRY C., ENGEL L.A.. Pressure-time product, flow, and oxygen cost of resistive breathing in humans. J. Appl. Phsyiol. 1985. 58: 1263-1272.
4. ELDRIDGE F.L., VAUGHN K.Z.. Relationship of thoracic volume and airway occlusion pressure: muscular effects. J. Appl. Physiol. 1977. 43-312-321.
5. EVANICH M.J., FRANCO M.J., LOURENCO R.V.. Force output of the diaphragm as a function of phrenic nerve firing rate and lung volume. J. Appl. Physiol. 1973. 35: 208-212.
6. FARKAS, G.A., ROUSSOS C.. Acute diaphragmatic shortening: in vitro mechanics and fatigue. Am. Rev. Respir. Dis. 1984. 130: 434-438.
7. KIM M.J., DRUZ W.S., DANON J., MACHNACH W., SHARP J.T.. Mechanics of the canine diaphragm. J. Appl. Physiol. 1976. 41: 369-382.
8. MACKLEM P.T., MACKLEM D.M., DE TROYER A.. A model of inspiratory muscle mechanics. J. Appl. Physiol. 1983. 55: 547-557.
9. MARSHALL R.. Relationship between stimulus and work of breathing at different lung volumes. J. Appl. Physiol. 1962. 17: 917-921.
10. MARTIN J.G., POWELL E., SHORE S., ENRICH J., ENGEL L.A.. The role of the respiratory muscles in the hyperinflation of bronchial asthma. Am. Rev. Respir. Dis. 1980. 121: 441-447.
11. MARTIN J.G., SHORE S., ENGEL L.A. Effect of continuous positive airway pressure on respiratory mechanics and pattern of breathing in induced asthma. Am. Rev. Respir. Dis. 1982. 126: 812-817.
12. MCCULLY K.K, FAULKNER J.A.. Length-tension relationship of mammalian diaphragm muscles. J. Appl. Physiol. 1983. 54: 1681-1686.
13. MINH V., DOLAN G.F., KONOPKA R.F., MOSER K.. Effect of hyperinflation on inspiratory function of the diaphragm. J. Appl. Physiol. 1976. 49: 67-73.
14. MULLER N., BRYAN A.C, ZAMEL N. Tonic inspiratory muscle activity as a cause of hyperinflation in histamine-induced asthma. J. Appl. Physiol 1980. 49: 869-874.
15. PENGELLY L.D., ALDERSON A.M., MILIC-EMILI J.. Mechanics of the diaphragm. J. Appl. Physiol. 1971. 30: 797-805.
16. ROUSSOS C., FIXLEY M., GROSS D., MACKLEM P.T.. Fatigue of the inspiratory muscles and their synergic behavior. J. Appl. Physiol. 1979. 46: 879-904.

A. Grassino - C. Fracchia - C. Rampulla - L. Zocchi
Respiratory Muscles in COPD
Bi & Gi Publishers, Verona - Springer-Verlag, London, 1988

Effect of lung volume on in vivo contraction characteristics of the human diaphragm

J. Smith, F. Bellemare
Meakins-Christie Laboratories, McGill University, Montreal, Quebec, Canada

SUMMARY: We performed transcutaneous bilateral phrenic nerve stimulation at varying lung volumes between RV and TLC in 6 normal male volunteers. Peak twitch transdiaphragmatic pressure declined from a mean of 49.1 (SD: 9.1) cm H_2O at RV to a mean of 19.6 cm H_2O (SD: 5.97) at TLC. Twitch contraction time fell from a mean of 91.8 (SD: 11.3) at RV to a mean of 57.7 ms (SD: 7.4) at TLC. There was a good correlation between changes in contraction time and transdiaphragmatic pressure (r = 0.7). The fall in transdiaphragmatic pressure was almost all due to a fall in pleural pressure, with little change in gastric pressure between RV and TLC. At TLC the pleural pressure in response to phrenic nerve stimulation was -0.58 cm H_2O. We conclude that as lung volume increases and the diaphragm shortens, it becomes less effective as a pressure. At a lung volume close to TLC the diaphragm ceases to act as an inspiratory muscle.

KEY WORDS:Lung volume; contraction characteristics, in vivo; diaphragm; phrenic nerve stimulation; transcutaneous bilateral; pleural pressure; gastric pressure; inspiratory muscle.

Introduction

The tension a muscle can generate is dependent on its initial length. In keeping with this it has been found that the maximal transdiaphragmatic pressure that the normal human diaphragm can generate varies inversely with lung volume. This has been demonstrated both during maximum voluntary effort[1, 4] and in response to unilateral transcutaneous phrenic nerve stimulation[19]. However, the use of maximum voluntary effort involves the coactivation of several other respiratory muscles, so that it is not possible with this method to study in isolation the contraction of the diaphragm. Furthermore, it is not always clear whether normal subjects are capable of fully activating their diaphragm voluntarily or whether such maximal activation can be maintained at different lung volumes. In our experience producing a maximal diaphragm contraction requires considerable motivation and practice and cannot always be achieved[2]. The problems associated with voluntary effort have been bypassed by Pengelly et al.[19], using transcutaneous stimulation of one phrenic nerve. That study provided the first in vivo measurement of the isolated contraction of the human diaphragm at different lung volumes. A recent study[3] pointed out, however, that unilateral stimulation in man may not provide an accurate representation of diaphragm contractility, the response being strongly conditioned by the compliance of the relaxed contralateral hemidiaphragm. With bilateral stimulation the normal geometry of the diaphragm is retained, and the transdiaphragmatic pressure response closely approximates to that determined in vitro.

Previous measurements using bilateral stimulation in man were restricted to resting end-expiratory lung volume[3]. The present study extends the analysis of the diaphragmatic twitch response in normal subjects to varying lung volumes encompassing the vital capacity.

Materials and methods

Six normal male subjects between the ages of 28 and 35 years were studied in sitting position. All subjects gave informed consent. They breathed through a Fleish 3 pneumotachograph to measure flow at the mouth. Lung volume changes were obtained by integration of the flow signal. Gastric pressure (Pga) and esophageal pressure (Ppl), which were taken as indices of abdominal and pleural pressure respectively, were measured using conventional balloon catheters and Sanborn 267BC differential pressure transducers. The balloons were positioned in the esophagus and stomach, as described by Milic-Emili et al.[16] and Agostoni et al.[1] respectively. Transdiaphragmatic pressure (Pdi) was obtained by electronically subtracting Ppl from Pga. Skin surface diaphragmatic EMG was recorded on each side of the chest using two pairs of 5 mm surface cup electrodes, applied to the skin over the sixth and seventh intercostal spaces near the costal margin. In 5 subjects the esophageal diaphragmatic EMG was recorded using an anchored electrode made from a modified Swan Ganz catheter mounted at the tip with two silver rings 20 cm apart as previously described[20]. The catheter was fixed at the gastro-esophageal junction by a 10 cc balloon filled with air. It was allowed to hang freely at the nares, and move with the gastroesophageal junction during respiration. The EMG signals were conditioned by 3 differential AC amplifiers (Bak Electronics, MDA-1). Phrenic nerve stimulation was performed using a Grass S88 stimulator and hand-held cathodes, as described by Bellemare and Bigland-Ritchie[2]. The terminal buttons were covered with gauze soaked in electroconductive cream. Aluminium patches strapped over the subclavicular area on each side of the chest were used as anodes. 100 us square wave pulses, at a rate of one per second and a constant current which could be varied between 0-50 ma, were applied to the phrenic nerve motor points in the neck, as described by Sarnoff et al.[23].

Inspiratory capacity, expiratory reserve volume and slow vital capacity were measured in all subjects. Stimulus maximality was then established for each phrenic nerve separately by applying single shocks while progressively increasing the stimulus current, until no further increase in ipsilateral skin surface mass action potential

(M-wave) was seen. These were continuously displayed and marked on the screen of a storage oscilloscope. This procedure was performed on each side separately at FRC, residual volume (RV), total lung capacity (TLC), half inspiratory capacity (1/2 IC) and half expiratory reserve volume (1/2 ERV), with the subjects relaxed against a closed shutter at each volume. For this purpose the volume signal was displayed on an oscilloscope placed in front of the subject. Occasionally skin surface electrode position which produced a good EMG signal at one lung volume would result in a very low signal amplitude at another lung volume. This may have been caused by the electrodes coming to overlie ribs as the lung volume changed. When this occurred the skin surface electrodes were changed in position to restore M-wave amplitude. At no time were the electrode positions changed after the establishment of stimulus maximality. The esophageal EMG was not used as a reference during the study, its position being dependent on the motion of the diaphragm. At each of the lung volumes, bilateral phrenic nerve simulation was performed, using current greater than that required to produce a maximal M-wave with unilateral stimulation.

All the pressures, as well as the mouth flow and volume signals, were recorded on a strip chart recorder. Mouthflow, Pdi, the 3 EMG signals and, in 4 subjects Pga were also recorded on tape. Pdi and EMG signals were analyzed from tape using an IBM XT computer. Signals were sampled at a rate of 1000 Hz. Ppl and Pga were analyzed from the strip chart recording. In two subjects baseline Ppl went off scale on the stripchart recorder at TLC, and was not included in the analysis. Phrenic nerve/conduction time was measured by playing back the esophageal EMG from tape to a storage oscilloscope, and directly measuring the time from stimulus to the start of the M-wave.

Only those twitches whose M-waves retained maximum amplitude, i.e. were equal to the control M-wave, were analyzed. At least 4-5 twitches in each individual at each lung volume were averaged, allowing the production of a composite twitch. For each individual twitch changes in Pdi, Pes and Pga were measured as peak amplitude above the immediately preceding baseline.

For each Pdi twitch we measured the contraction time (CT) and half relaxation time (1/2 RT), defined as the time from onset to peak pressure and from peak to half peak pressure decay respectively. The maximal contraction rate (MCR) and maximal relaxation rate (MRR) were measured by differentiating the Pdi waveform. MCR and MRR were normalized by dividing by the twitch peak Pdi. The time constant of the late part of the twitch relaxation, when monoexponential, was also measured.

Differences between parameters at different lung volumes were analyzed using a split plot analysis of variance for repeated measures and Scheffes test. The correlation between PDi and CT was assessed by calculating the product moment correlation. All statistical analysis was performed on an IBM XT computer using the Statistics package (NH analytical software).

Results

Vital statistics of the subjects are shown in Table 1. As shown in Table 2, the mean test skin surface M-waves on the left and right sides during bilateral stimulation were in general of greater amplitude than the corresponding control values obtained with unilateral stimulation. Since the position of the skin surface electrodes was often changed between lung volumes, no comment can be made as to their relative amplitude at different lung volumes. The esophageal M-wave increased consistently from a mean of 2.8 (SD: 2.25) mvolts at RV to 8.12 (SD: 3.03) mvolts at TLC. Therefore, the esophageal M-wave could not be used to establish the maintenance of stimulus maximality at different lung volumes. The amplitude and waveform of the Pdi twitches at any lung volume in any one subject were highly reproducible. An example of 5 superimposed Pdi twitches from one subject at residual volume is shown in Figure 1A. Composite twitches from the same subject at all five lung volumes are shown in Figure 1B. The amplitude and duration of the twitches decreased between RV and TLC. These changes occurred consistently in all subjects. The mean values and standard deviations of peak Pdi, contraction time and 1/2 relaxation time for each subject are shown in Table 3, and the corresponding mean group values are graphed in Figs. 3, 4, and 5. As can be seen,

Table 1. - Vital statistics and lung volumes of each subject. VC: vital capacity; IC: inspiratory capacity; ERV: expiratory reserve volume; HGT: height; WGT: weight.

Subject	Age (years)	VC (1)	IC (1)	ERV (1)	HGT (cm)	WGT (kg)
1	35	5.6	2.9	2.7	168	60.5
2	35	4.7	3.2	1.5	168	65.0
3	28	4.0	2.8	1.2	167	65.0
4	30	3.8	2.6	1.2	168	65.0
5	28	5.2	3.2	2.0	169	69.1
6	28	3.6	2.4	1.2	167	51.6

Table 2. - Skin surface mass action potential (M-wave) amplitude from right and left hemidiaphragms during bilateral stimulation at different lung volumes. TLC: total lung capacity; 1/2 IC: half inspiratory capacity; FRC: function residual capacity; 1/2 ERV: half expiratory reserve volume; RV: residual volume. The amplitude is expressed as a fraction of the control maximum amplitude obtained at that volume using unilateral stimulation. The mean for six subjects is given. The numbers in brackets indicate standard deviation.

Volume	Right m-wave	Left m-wave
TLC	1,54 (0.67)	1.21 (0.23)
1/2 IC	1.90 (0.86)	1.30 (0.35)
FRC	1.11 (038)	1.12 (0.58)
1/2 ERV	1.16 (0.24)	1.34 (0.54)
RV	1.28 (0.34)	0.99 (0.65)

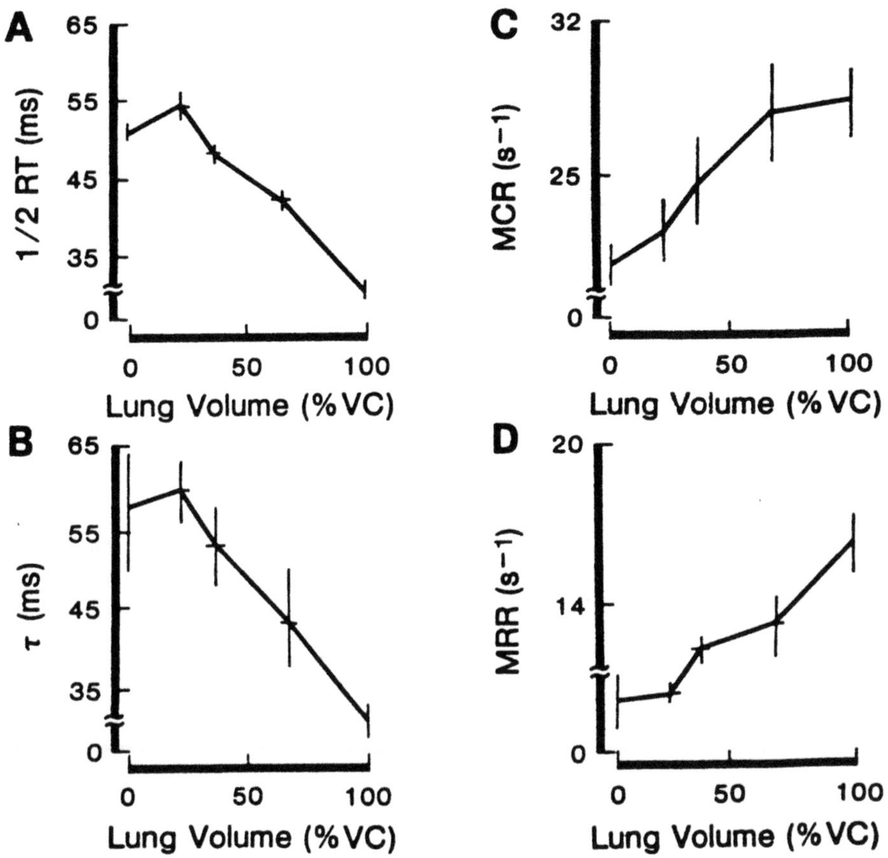

Fig. 1

Pdi increased from a mean of 19.59 cm H_2O at TLC to a mean of 49.14 cm H_2O at RV. Pressure differences between each of the lung volumes were statistically significant ($p<0.01$) and in all but one subject, the highest twitch Pdi was recorded at RV. Contraction time consistently increased between TLC and RV, though the difference between the values at 1/2 ERV and RV was not statistically significant.

Figure 5 shows the changes in 1/2 RT, MCR, MRR and time constant of relaxation. As can be seen, changes in all parameters were indicative of a decreased twitch duration with increased lung volume. However, the changes in these parameters were not as consistent as the changes in CT and were of statistical significance only between some volumes. As shown in Figure 6, a significant correlation was found between CT and peak twitch Pdi ($r = 0.70$).

As the skin surface electrodes had been changed in position between lung volumes, the skin surface M-waves could not be used to measure phrenic nerve conduction time. In three subjects the start of the M-wave could be clearly delineated on the esophageal EMG and no change in phrenic nerve conduction time was seen with

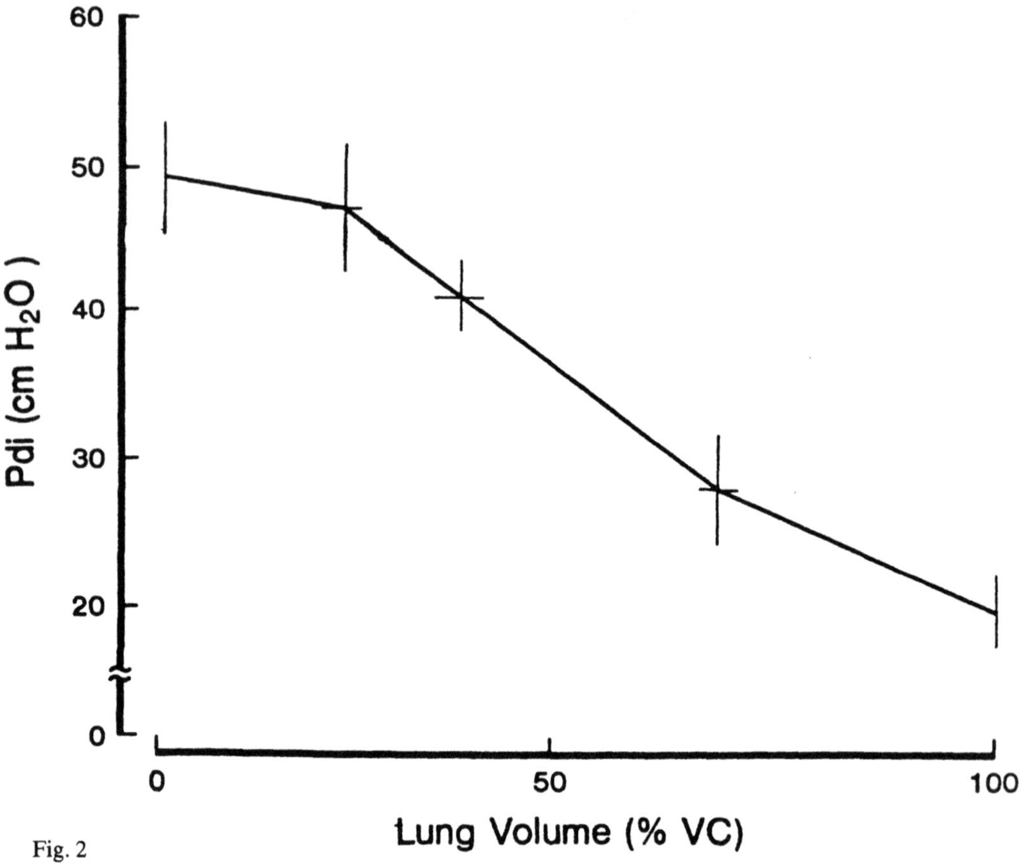

Fig. 2

changes in lung volume in these subjects. Time from stimulus to onset of contraction (latent period) increased consistently in all subjects from a mean of 16 ms (SD: 3.6) at RV to a mean of 22 ms (SD: 2.4) at TLC (Fig. 1B).

As lung volume increased from RV to TLC the contribution of Pga and Ppl to Pdi changed (Fig. 6). Pga tended to remain constant at all lung volumes, whereas Ppl fell progressively between RV and TLC. At TLC the mean Ppl was -0.58 cm H_2O. If we estimate the Ppl for the two subjects whose baseline Ppl went offscale at TLC (by subtracting Pga from Pdi), the mean Ppl at TLC would be -2 cm H_2O. One subject (subject 6) developed a positive pressure of 8.2 cm H_2O in response to BPS at TLC, i.e., his diaphragm appeared to act as an expiratory muscle at TLC.

Discussion

The contractile response of the human diaphragm has been evaluated using maximal bilateral phrenic nerve stimulation at different lung volumes. Such determinations have not been reported previously. The results showed: 1) that Pdi in response

Table 3. - Mean twitch transdiaphragmatic pressure (Pdi), contraction time (CT), and half relaxation time (1/2 RT), for each subject separately and for the group, at each lung volume. Volumes and other conventions as in Table 2.

Subject	TLC	1/2 RC	FRC	1/2 ERV	Rv
		Pdi			
1	26.8 (2.05)	35.8 (1.35)	46.1 (1.16)	44.4 (3.12)	46.1 (1.25)
2	18.8 (0.27)	31.4 (0.99)	38.1 (3.04)	58.8 (3.71)	57.4 (1.25)
3	17.4 (1.03)	18.9 (0.76)	40.1 (1.29)	58.1 (2.54)	61.0 (1.95)
4	12.2 (1.49)	16.9 (1.06)	40.7 (1.46)	36.0 (1.52)	40.3 (1.22)
5	15.6 (0.68)	25.1 (1.58)	33.3 (1.84)	34.4 (1.56)	38.5 (2.22)
6	26.7 (2.02)	39.0 (0.81)	47.0 (1.24)	50.8 (3.38)	51.5 (1.75)
Group mean	19.6 (5.97)	27.8 (9.03)	40.89 (5.1)	47.1 (10.61)	49.1 (9.1)
		CT			
1	59.3 (3.1)	69.5 (1.1)	69 (3.4)	81 (7.8)	94.5 (1.3)
2	60 (3.4)	78 (2.2)	82 (1.1)	101 (3.7)	99 (4.8)
3	46 (1.8)	64 (3.4)	87 (1.3)	90 (5.0)	101 (2.1)
4	52 (10.9)	66 (3.4)	75 (2.4)	74.5 (7.2)	81.5 (2.5)
5	67.5 (3.5)	78 (2.5)	88.5 (3.1)	96 (14.4)	100 (2.4)
6	61 (2.6)	68 (3.3)	70 (3.1)	88 (2.5)	74 (2.4)
Group mean	57.7 (7.4)	70.5 (6.1)	79.6 (9.5)	88.4 (9.7)	91.8 (11.3)
		1/2 RT			
1	41 (2.0)	58.5 (4.7)	59.5 (3.7)	65.5 (5.9)	76 (2.9)
2	44 (6.7)	42 (3.4)	54 (3.1)	59 (1.8)	75 (2.6)
3	42 (3.5)	58 (4.3)	56 (2.2)	61.5 (1.9)	69 (4.0)
4	55 (8.7)	59 (11.0)	57 (3.1)	78 (7.0)	82 (5.0)
5	46 (1.4)	65 (6.1)	57.5 (4.8)	55.6 (10.6)	56 (6.6)
6	30 (2.3)	42 (6.1)	48 (3.1)	54 (5.3)	51 (2.9)
Group mean	43.0 (8.2)	54.3 (9.9)	55.4 (4)	62.6 (9.2)	68.2 (12.3)

to constant bilateral maximal nerve stimuli fell with increasing lung volume. This is consistent with the diaphragm contracting at a progressively shorter initial fiber length. The highest Pdi twitch amplitude was recorded at RV, suggesting that the optimal length of the diaphragm lies at or below RV; 2) at TLC the diaphragm was still capable of generating considerable tension, i.e., about 40% of the maximum recorded at RV. However, it was unable to convert this tension into useful inspiratory pleural pressure; 3) in accordance with in vitro studies the duration of single Pdi twitches decreased systematically with lung inflation, suggesting that progressively higher rates of stimulation would be required to achieve a maximal diaphragmatic contraction as lung volume increases.

Length-Tension Property

Although the observed fall in twitch Pdi with increasing lung volume can be explained on the basis of length-tension properties, other factors may also be involved. An increased radius of curvature at high lung volumes could alter the pressure output of the diaphragm, irrespective of its length-tension properties[15]. Further-

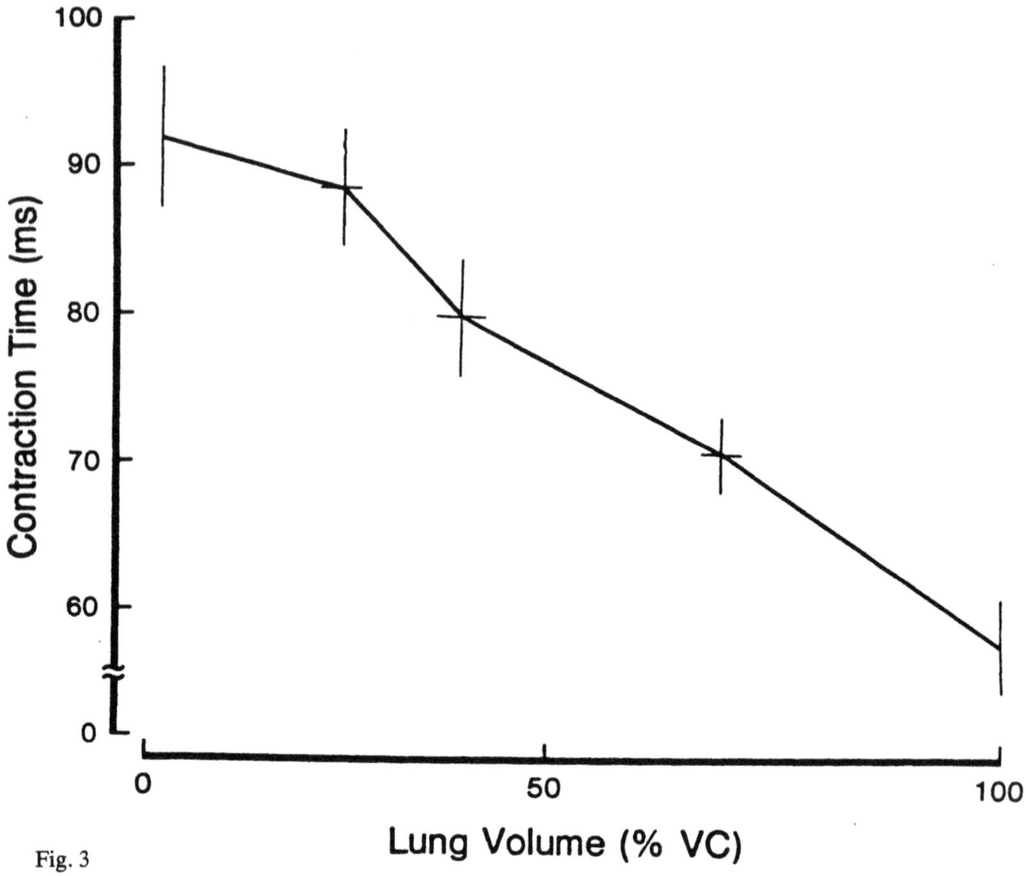

Fig. 3

more, in our study single twitches were employed, while in in-vitro studies the length-tension properties are generally determined using tetanic stimulation. In humans, it has not so far been found possible to stimulate the phrenic nerves bilaterally at a frequency greater than 35 Hz[3]. This frequency is considerably less than that required to produce a maximal contraction. Limb[21] and in-vitro diaphragm[10] studies have shown a decrease in the twitch to tetanus tension ratio as muscle shortens below its optimal length. Hence both an increase in radius of curvature and a decrease in twitch to tetanus tension ratio could in our study lead to a progressive underestimation of the maximum available diaphragm tension as lung volume increases.

It is thus desirable to compare our Pdi twitch measurements with the tension/length relationship of the diaphragm determined in vitro using tetanic stimulation. The relationship between lung volume and diaphragm length has been studied in man by Braun et al.[4] and Loring et al.[13]. Both studies indicated that the diaphragm is about 65% shorter at TLC than at RV. Their data can therefore be used to convert the lung volume at which our Pdi measurements were made to equivalent diaphragm length changes.

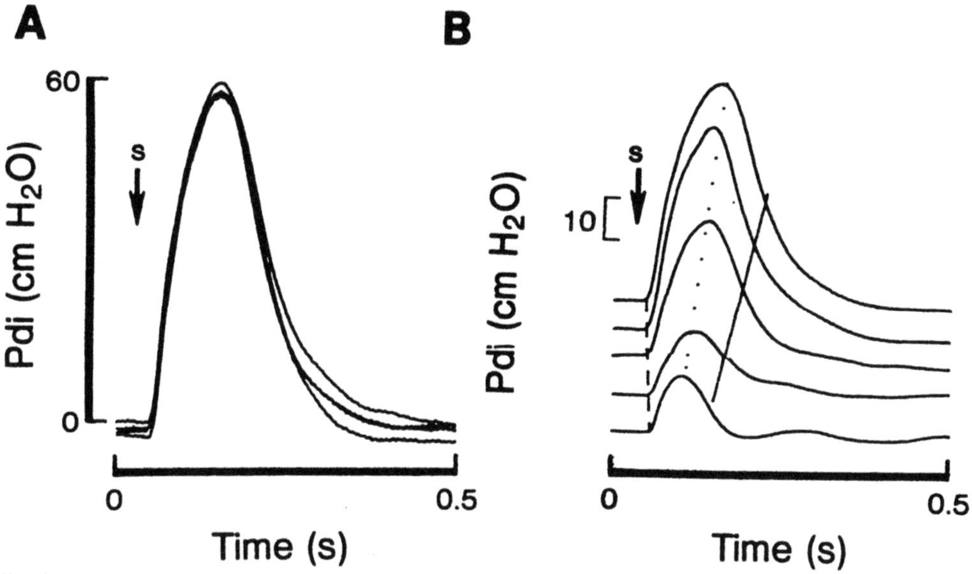

A
B
60
0
s
Pdi (cm H_2O)
0
0.5
Time (s)
s
10
Pdi (cm H_2O)
0
0.5
Time (s)

Fig. 4

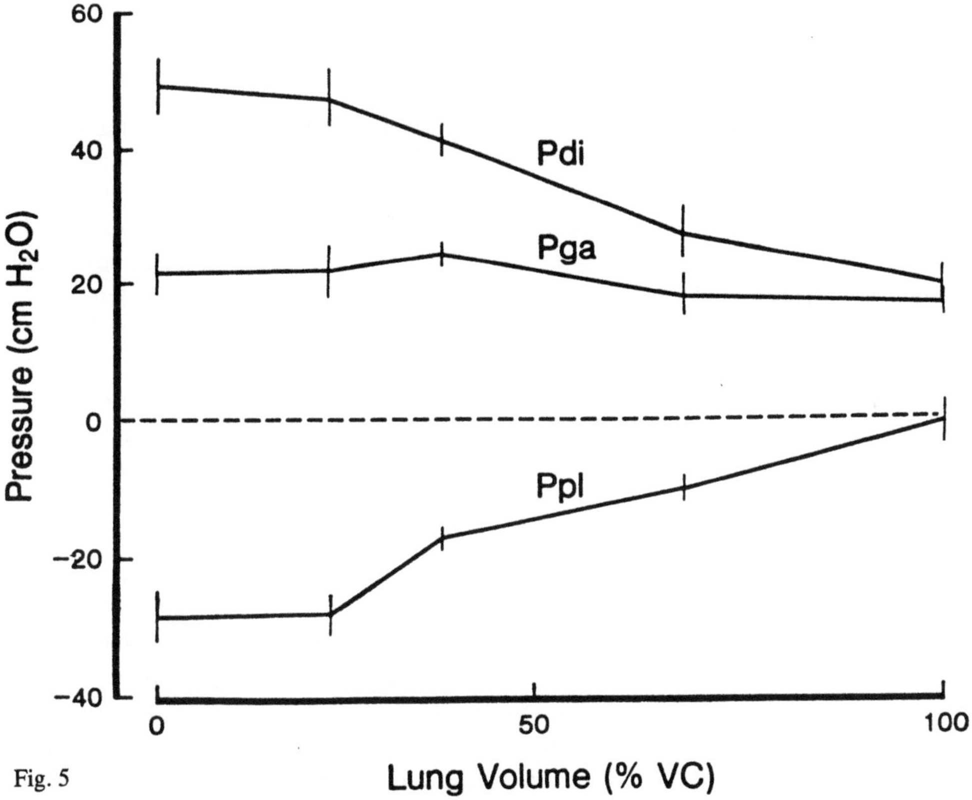

60
40
20
0
−20
−40
Pressure (cm H_2O)
Pdi
Pga
Ppl
0
50
100
Lung Volume (% VC)

Fig. 5

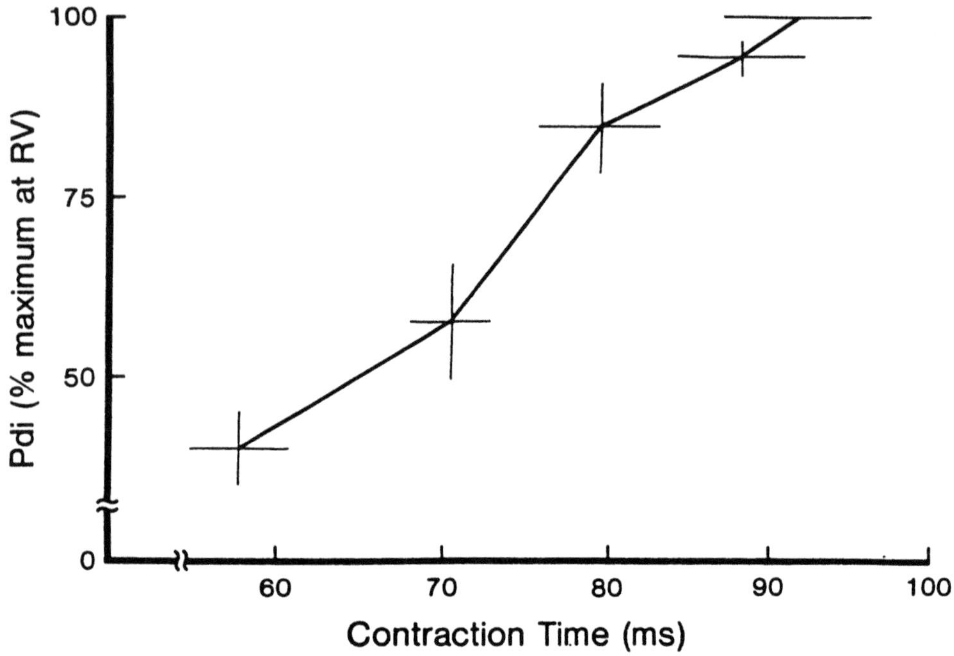

Fig. 6

In Figure 7 the relationship thus obtained between the Pdi twitch amplitude (expressed as percent of maximum RV value) and the estimated equivalent diaphragm length is compared to the tension-length relationship, reported by McCully and Faulkner[14] for the isolated diaphragm of different species and by Edwards and Faulkner[8] for isolated human diaphragm strips studied in vitro. Since in both our study and that of Braun et al.[4] the highest Pdi were recorded at RV, the length of the diaphragm at that volume is assumed to be the optimal length of the diaphragm (100% in Figure 7). As can be seen, our in vivo data using single twitches is in close agreement with the in vitro length-tension relationship reported by other workers. This comparison thus suggests that the observed decline in Pdi twitch amplitude with increasing lung volume can be accounted for on the basis of the length/tension properties of the diaphragm alone. This in turn would imply that the radius of curvature of the diaphragm did not change appreciably within the range of lung volume studied or that such changes had little influence on the pressure developed by the diaphragm at different lung volumes. This conclusion is consistent with the studies of Braun et al.[4] and Loring et al.[13], showing little if any change in the shape of the human diaphragm between RV and TLC. Furthermore, according to Figure 7, the optimal length of the human diaphragm would lie at a lung volume close to RV.

The similarity of our single twitch data with the in vitro tetanic length/tension relationship of the diaphragm is nevertheless surprising. The twitch amplitude is

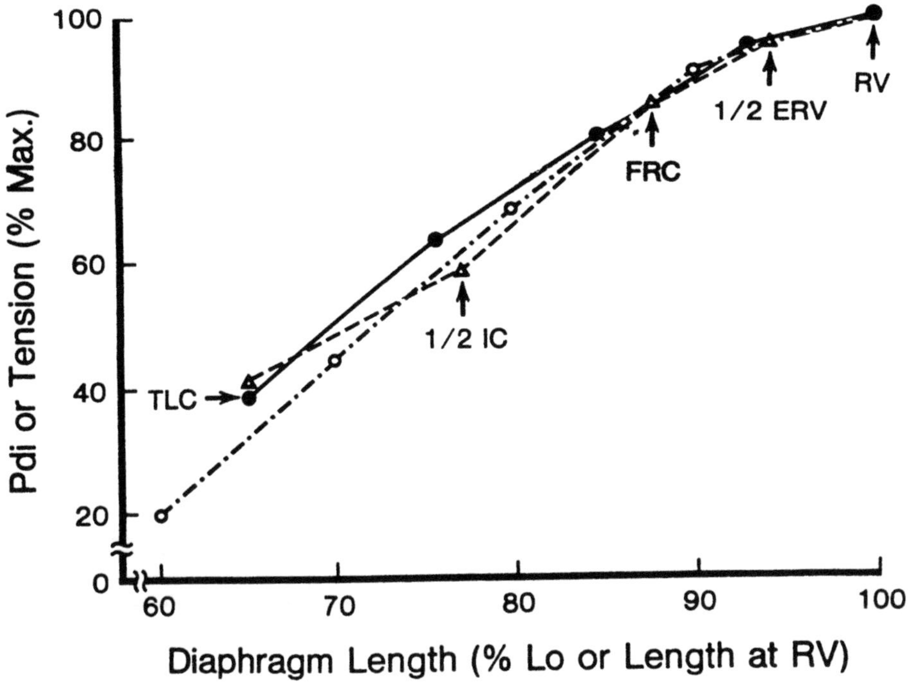

Fig. 7

determined both by the rate of rise of pressure or force and by the twitch contraction time. As for isolated muscle experiments[10, 21], we found that CT decreased consistently with shortening of the diaphragm, and a nearly linear relationship was found between Pdi twitch amplitude and CT. Since the rate of rise of pressure or force would be expected to fall in proportion to the maximum tetanic tension, a decrease in CT should cause the final amplitude to be smaller and the twitch to tetanus tension ratio to fall. As shown in Figure 8, the rate of rise of Pdi at high lung volumes was considerably above that predicted on the basis of the changes in tetanic tension. It thus appears that in our study the effect of diaphragm shortening on CT was compensated by a relative increase in the rate of rise of Pdi. The mechanism underlying this finding is unclear. When recorded in vitro, the rate of rise of force in a single twitch will be determined by the velocity of shortening of the muscle fibers and the compliance of the series elastic elements[14, 21]. Since the diaphragm in vivo is also in series with the chest wall and lungs, the rate of rise of Pdi will also be determined by the compliance of these structures. The decrease in compliance of the lungs, rib cage and abdomen as lung volume increases could thus account for the relative increase in the rate or rise of Pdi. Whatever the mechanism, this increased rate of rise is a useful coincidence; for the comparison shown in Figure 7 suggests that changes in maximum diaphragm strength at different lung volumes

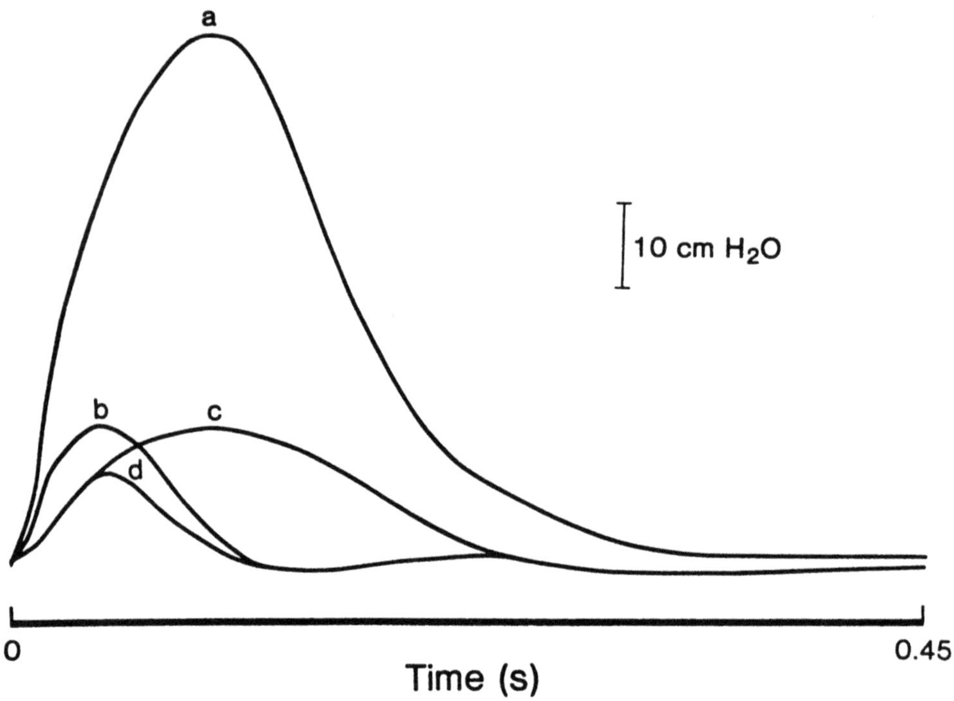

Fig. 8

can be reasonably estimated by recording single bilateral diaphragmatic twitches from the relaxed muscle.

Despite the apparent increase in rate of rise of Pdi, the fall in twitch CT with increasing lung volume may still have some implications; for higher excitation rates would be required to achieve the same degree of mechanical fusion in a muscle with a short twitch duration. Edwards has indeed shown previously, for isolated human diaphragm strips in vitro, a linear relationship between the 20 to 100 Hz tension ratio (an index of mechanical fusion) and diaphragm length[7]. Hence higher phrenic motor neuron discharge rates would be required to achieve any given fraction of the maximum available force as lung volume increases.

Inspiratory Function of the Diaphragm

The fall in Pdi with increasing lung volume was almost solely due to a fall in Ppl, while Pga remained unchanged. This is consistent with previous studies in cats[15, 19], rabbits[5, 22], and dogs[5], showing that the effectiveness of the diaphragm as an inspiratory pressure generator decreases with increasing lung volume. Furthermore, at TLC, as in rabbits[5, 22] and dogs[5], Ppl was virtually nil, indicating that in all species the diaphragm ceases to act as an inspiratory pressure generator when the lung volume is close to TLC. At higher lung volumes, as in one of our subjects at TLC, the diaphragm may even act as an expiratory muscle[17, 22].

However, a major difference between our results and the above-mentioned studies is the fact that the human diaphragm at TLC is still capable of generating considerable tension (about 40% of the maximum recorded at RV), but is no longer able to convert this tension into useful inspiratory pressure. Thus, in species other than man, Ppl falls in proportion to Pdi and the diaphragm ceases to act as an inspiratory pressure generator when its fibers are maximally shortened and can no longer develop active tension. This was verified in supine dogs by Kim and associates[12], who showed that the maximal tangential diaphragm tension measured directly in response to bilateral tetanic phrenic nerve stimulation was zero at a lung volume close to TLC.

Previous animal studies were performed in supine position. However, results similar to ours were reported by Danon et al.[6] in one supine tetraplegic patient whose phrenic nerves were paced bilaterally at a lung volume close to TLC. Thus the fact that our results were obtained in a seated posture is unlikely to account for the difference observed between humans and other species. The different behavior of the human diaphragm at high lung volumes is unlikely to be explained by interspecies differences in the active length/tension properties, since this relationship is remarkably similar among all species[14]. Equally, it cannot be explained by a different amount of passive shortening with hyperinflation, since this also appears to be similar in different species. Indeed, the length of the human diaphragm at relaxed TLC was estimated to be between 55 and 70% of its optimal length at RV[4, 13] (see also Figure 7). In supine dogs, the passive diaphragm at TLC is about 30-35% shorter than at FRC[18]. Since the optimal length of the diaphragm in supine dogs is about 10% greater than at FRC[12], the length of the dog diaphragm at relaxed TLC is also estimated to be between 55 and 60% of its optimal length.

The difference in Pdi developed at high lung volumes in humans as compared to other species is more likely to be related to a different amount of active shortening, as a result of chest wall distortions introduced by the isolated contraction of the diaphragm.

Several detailed studies have described the chest wall distortions produced by the action of the diaphragm when the airways are closed. These vary both qualitatively and quantitatively among species, as well as with changes in lung volume[5, 22]. The general view is that at low lung volumes the fall in intrathoracic pressure draws the upper rib inward, with a consequent descent of the diaphragm dome and increase in abdominal pressure. This action, being uniquely dependent on the fall in intrathoracic pressure, vanishes as the lung volume increases or as the intrathoracic pressure swing decreases[5]. At high lung volume, the action of the diaphragm tends to draw the lower rib cage inward. In rabbits this occurs at all lung volumes[5, 22], while in dogs it is only observed at volumes greater than 70% TLC[5]. This latter action is largely determined by the orientation of the diaphragmatic muscle fibers relative to the rib cage, and becomes more prominent as lung volume increases. The

Table 4. - Changes in gastric (ΔPga) and pleural (ΔPpl) pressures and the ratio $\Delta ga/\Delta Ppl$ in different species during bilateral phrenic nerve stimulation at end-expiratory lung volume (FRC) and at a volume close to total lung capacity (TLC) when the airways are closed, calculated from the references indicated in brackets.

	Humans		Dogs	Cats	Rabbits
	supine[1,6]	seated[1,6]	supine[5]	supine[15,19]	supine[5,22]
		FRC			
ΔPga (cmH_2O)	8.0	24.1	5.3	1.8	2.0
ΔPpl (cmH_2O)	14.0	17.1	30.1	12.8	30.9
$\Delta Pga/\Delta Ppl$	0.57	1.41	0.18	0.14	0.07
		Near TLC			
ΔPga (cmH_20)	11.5	16.7	2.0	–	0.6
ΔPpl (cmH_20)	5.0	2.0	3.0	–	0.5
$\Delta Pga/\Delta Ppl$	2.3	8.4	0.67	–	1.2

common feature of all these distortions is a shortening of the diaphragm as it contracts. Thus the isolated contraction of the diaphragm in vivo is far from being isometric. In other words, the contraction of the diaphragm in vivo is truly isometric only at the point of maximal distortions, which in turn must correspond to the point of maximum pressure changes: in our study, at the peak of the twitch. It is easy to understand that the final isometric tension or Pdi thus measured at any given lung volume will be dependent on the extent of active diaphragm shortening and hence on the magnitude of chest wall distortions. Since the latter are conditioned by the elastance of the chest wall, and since the chest wall elastance increases with body size among species[11], the better maintained Pdi at different lung volumes in humans is probably related to their stiffer chest wall. Previous studies did not indicate whether the abdomen or the rib cage is mainly responsible for the stiffer chest wall in large as compared to smaller animals. However, the elastance of the abdomen (Eab) relative to that of the rib cage (Erc) can be roughly estimated from the ratio of the changes in abdominal (ΔPga) and pleural (ΔPpl) pressure during artificial phrenic nerve stimulation at iso-lung volume. This is possible since, under isovolume conditions, the volume changes of the abdominal (Vab) and rib cage (Vrc) compartments are equal and opposite:

and $Vab = -Vrc = \Delta Pab/Eab = \Delta Ppl/Erc$[1]

thus $$\frac{Eab}{Erc} = \frac{\Delta Pab}{\Delta Ppl}$$

The ratio $\Delta Pab/\Delta Ppl$ has been computed from this and other studies of diaphragm response to bilateral phrenic nerve stimulation in different species[5, 6, 19]. The results obtained at FRC and at a lung volume close to TLC are summarized in Table 4. As can be seen, the elastance of the abdomen occupies a progressively greater proportion of the total elastance in large as compared to smaller animals. Furthermore, the abdominal elastance increases markedly in all species between FRC and TLC. In rabbits, cats and dogs, however, the abdominal elastance appears to be a relatively small fraction of the total elastance. In these species therefore, the extent of active diaphragm shortening should be largely determined by the elastance of the rib cage. In marked contrast, in seated humans at FRC and seated or supine humans at TLC, the abdominal elastance appears to exceed markedly that of the rib cage. The stiffer abdominal wall therefore, while maintaining the diaphragm at a more suitable optimal length at all lung volumes, may also be the main factor limiting the inspiratory function of the human diaphragm.

In conclusion, our results have shown that the maximal tension the diaphragm can generate decreases with increasing lung volume in a way predicted by its length/tension properties. However, the results also suggest that the mechanical coupling of the diaphragm with the chest wall is a far more important determinant of the inspiratory function of the diaphragm than has hitherto been assumed.

Acknowledgements

This study was supported by the Medical Research Council of Canada and the Parker B. Francis Foundation.

References

1. AGOSTONI E., RAHN H.: Abdominal and thoracic pressures at different lung volumes. J. Appl. Physiol. 1960. 1087-1092.
2. BELLEMARE F., BIGLAND-RITCHIE B.: Assessment of human diaphragm strength and activation using phrenic nerve stimulation. Respir. Physiol. 1984. 58: 263-277.
3. BELLEMARE F., BIGLAND-RITCHIE B., WOODS J.J.: Contractile properties of the human diaphragm in vivo. J. Appl. Physiol. (In Press).
4. BRAUN N.M., ARORA N.S., ROCHESTER D.F.: Force length relationship of the normal human diaphragm. J. Appl. Physiol. 1982. 53: 405-412.
5. D'ANGELO E., SANT'AMBROGIO G.: Direct action of contracting diaphragm on the rib cage in rabbits and dogs. J. Appl. Physiol. 1974. 36: 715-719.
6. DANON J., DRUZ W.S., GOLDBERG N.B., SHARP J.T.: Function of the isolated paced diaphragm and the cervical accessory muscles in C1 quadriplegics. Am. Rev. Respir. Dis. 1979. 119: 909-919.
7. EDWARDS R.H.T.: The diaphragm as a muscle: Mechanism underlying fatigue. Am. Rev. Respir. Dis. 1979. 119 (suppl. part 2): 81-84.
8. EDWARDS R.H.T., FAULKNER J.A.; Lung biology in health and disease, In: The Thorax, Part A. Respiratory muscles: Structure and Function. New York, Dekker, 1986. Vol. 29, p. 297.

9. EVANICH M.I., FRANCO M.J., LOURENCO R.V. Force output of the diaphragm as a function of phrenic nerve firing rate and lung volume. J. Appl. Physiol. 1973. 35: 208-212.
10. FARKAS G.A., ROUSSOS CH.: Acute diaphragmatic shortening: In vitro mechanics and fatigue. Am. Rev. Respir. Dis. 1984. 130: 434-438.
11. GILLESPIE J.L.: Mechanisms that determine functional residual capacity in different mammalian species. Am. Rev. Respir. Dis. 1983. 128: (suppl. part. II): S74-S77.
12. KIM M.J., DRUZ W.S., DANON J., MACHNACK W., SHARP J.T.: Mechanics of the canine diaphragm. J. Appl. Physiol. 1976. 41: 369-382.
13. LORING S.H., MEAD J., GRISCOM N.T.. Dependance of diaphragmatic length on lung volume and thoracoabdominal configuration. J. Appl. Physiol. 1985. 59: 1961-1970.
14. MCCULLY K.K., FAULKNER J.A.: Length tension of mammalian diaphragmatic muscles. J. Appl. Physiol. 1983. 54: 1681-1686.
15. MARSHALL R.: Relationships between stimulus and work of breathing at different lung volumes. J. Appl. Physiol. 1962. 17: 917-921.
16. MILIC-EMILI J., TURNER J.M., GLAUSER E.M.: Improved technique for estimating pleural pressure from esophageal balloons. J. Appl. Physiol. 1964. 19: 207-211/
17. MINH V.D., DOLAN G.F., KANOPKA R.F., MOSER K.M.: Effect of hyperinflation on inspiratory function of the diaphragm. J. Appl. Physiol. 1976. 40: 67-73.
18. NEWMAN S., ROAD J., BELLEMARE F., CLOZEL J.P., LAVIGNE C.M., GRASSINO A.: Respiratory muscle length measured by sonomicrometry. J. Appl. Physiol. 1984. 56 753-764.
19. PENGELLY M.D., ALDERSON A., MILIC-EMILI J.: Mechanics of the diaphragm. J. Appl. Physiol. 1971. 30: 797-805.
20. PETIT J.M., MILIC-EMILI J., DELHEZ L.: Role of the diaphragm in breathing in conscious man, an electromyography study. J. Appl. Physiol. 1964. 15: 1101-1106.
21. RACK P.M.H., WESTBURY D.R.: The effects of length and stimulus rate on tension in the isometric cat soleus muscle. J. Physiol. (London) 1969. 204: 443-460.
22. SANT'AMBROGIO G., SAIBENE F.: Contractile properties of the diaphragm in some mammals. Resp. Physiol. 1970. 10: 349-357.
23. SARNOFF S.J., SARNOFF L.C., WHITTENBERG J.L.: Electrophrenic respiration. VII. The motor point of the phrenic nerve in relation to external stimulation. Surg. Gynecol. Obstet. 1951. 93: 190-196.

A. GRASSINO - C. FRACCHIA - C. RAMPULLA - L. ZOCCHI
Respiratory Muscles in COPD
Bi & Gi Publishers, Verona - Springer-Verlag, London, 1988

Nutritional and metabolic aspects of COPD

E. FIACCADORI, S. DEL CANALE, A. GUARIGLIA.
Institute of Clinical Medicine and Nephreology, University of Parma, Italy

SUMMARY: Derangements of cell metabolism in course of hypercapnic-hypoxemic Chronic Obstructive Pulmonary Disease (COPD) are not well defined. This seems to be surprising, should one consider that the understanding of the response of body tissues to blood gas alterations is of utmost importance in order to assess the systemic effects of the disease.

The present paper is thus aimed at giving information on cell metabolism derangements in COPD patients with hypercapnia and hypoxemia. In particular, intracellular acid-base, electrolyte and energy metabolism in skeletal muscle will be analyzed.

Moreover, since some of the features of skeletal muscle cell metabolism presented here can be related to the nutritional status of COPD patients, preliminary data concerning a cross-sectional nutritional study on COPD patients will also be presented.

KEY WORDS: Nutritional aspects; COPD; metabolic aspects; hypercapnia; hypoxemia; skeletal muscle; nutritional status.

Alterations of intracellular acid-base and electrolyte metabolism are well known as factors negatively affecting tissue and organ functions. Nevertheless, only a few studies on intracellular acid-base and electrolyte metabolism in COPD patients are currently available.

No changes in the whole body pH (DMO method) or in muscle K^+ and Na^+ were found in a 1970 study performed on patients with moderate degree COPD[1]. In a further study on a series of COPD patients with chronic cor pulmonale, total body K^+ proved to be decreased[2]. According to a more recent study on quadriceps femoris muscle, muscle K^+ was unchanged while muscle Mg^{2+} was reduced[3]. Later on, our group presented the results of a study performed on quadriceps femoris skeletal muscle of patients with COPD and Acute Respiratory Failure (ARF)[4]. Muscle samples were analyzed in order to obtain:
- intracellular bicarbonate (HCO_3^{2-}) and pH (pHi) by means of total acid-labile carbon-dioxide (TCO_2) method[5];
- total muscle water (H_2Om) and muscle water compartments extracellular (H_2Oe) and intracellular water (H_2Oi)[5];
- muscle electrolytes: sodium (Na^+m), potassium (K^+m) and magnesium ($Mg^{2+}m$)[6].

An intracellular acidosis was present in the skeletal muscle of COPD patients with ARF (Tab. 1). pHi was significantly decreased in spite of intracellular bicarbonate increase. An increase of total muscle water, extracellular water, muscle sodium and chloride content was also found. Muscle potassium and magnesium content was reduced.

In COPD patients a close relationship between intracellular acidosis and arterial carbon-dioxide partial pressure was found (Fig. 1), though not between intracellular and extracellular pH.

The calculations of intracellular bicarbonate and pH from TCO_2, as well as of water compartments and intracellular electrolytes, are influenced by the membrane potential (Em) values used. No data on Em values in COPD patients are reported in literature, nor was this parameter measured in our patients. Therefore, values of both 88.8 mv (mean Em value measured in normal subjects) and 66.3 mv (mean Em value measured in a group of severely ill patients)[7] were assumed for COPD patients. In Table 2, water compartments, intracellular acid-base and electrolyte values of COPD patients, all calculated assuming different membrane potential values, are shown: the value of pHi is significantly reduced, whatever the Em values assumed for calculation. Moreover, muscle potassium content was reduced even when reference was made to intracellular water content, whatever the membrane potential values assumed for calculation.

Muscle cell potassium reduction was related neither to extra- or intracellular acid-base equilibrium parameters nor to extracellular potassium values. A significant relationship was found between muscle potassium and magnesium contents (Fig. 2).

As for the intracellular acid-base equilibrium, the following points can be made:

- A condition of intracellular acidosis was present in the skeletal muscle of the COPD patients with ARF.
- Intracellular acidosis was related to $PaCO_2$; as a result, skeletal muscle intracellular pH appears to be closely related to a systemic index of CO_2 retention;
- Intracellular acidosis was present even at normal extracellular pH values. No correlation was present between extra- and intracellular pH values. This could mean that the effects of the CO_2 retention on the intracellular compartment cannot be predicted from extracellular pH measurement alone.

As for the intracellular electrolyte metabolism, the following points on potassium depletion in COPD patients with ARF should be underlined:

- A condition of potassium depletion (i.e. cell potassium values lower than two standard deviations from the control mean) is detectable in about 30 of the COPD patients considered in this study.
- Potassium depletion was unrelated to extracellular potassium values and was also present at normal extracellular potassium values.
- Potassium depletion was unrelated to intracellular or extracellular acid-base equilibrium parameters, but related to muscle Mg content decrease.

Table 1 - Intracellular acid-base values, water compartments and electrolytes in muscle biopsy samples from the quadriceps femoris in the COPD patients group and in the control subjects group. Results are presented as mean values ± SEM. Significance of differences versus control subjects:
* $P < 0.001$, ° $P < 0.01$, + $P < 0.05$.

	Acid-labile TCO_2 mmol/kg muscle w.w.	$(HCO_3)_i$ mmol/l $(H_2O)_i$	pH_i	$(H_2O)_m$	$(H_2O)_e$	$(H_2O)_i$	Cl^-_m	NA^+_m	K^+_m	Mg^{2+}_m
				l/kg FFS			mmol/kg FFS			
COPD patients n = 21	18.5* ±0.7	13.9+ ±0.7	6.77* ±0.04	3.86* ±0.08	0.84* ±0.05	3.00 ±0.10	108* ±4	181 ±6 (n = 20)	400* ±8 (n = 20)	39° ±0 (n = 20)
Control subjects n = 21	12.2 ±0.2	12.3 ±0.3	7.00 ±0.01	3.45 ±0.06	0.60 ±0.02	2.86 ±0.05	81 ±2	102 ±3 (n = 19)	445 ±5 (n = 19)	44 ±0 (n = 19)

Table 2 - Water compartments, intracellular acid-base and electrolytes values in muscle biopsy samples from the quadriceps femoris in the COPD patients group and in the control subjects group, calculated assuming different membrane potential values. Results are presented as mean values ± SEM. Significance of differences versus control subjects:
* $P < 0.001$, + $P < 0.05$

	E_m	$(H_2O)_m$	$(H_20)_e$	$(H_2O)_i$	K_i	Na_i	$HCO_3{}^-_i$	pH_i
	mV	l/kg FFS			mmol/l $(H_2O)_i$			
COPD patients	−88.8	3.86* ±0.08	0.99* ±0.05	2.87 ±0.10	139.3 ±5.5+ (n = 20)	10.6 ±1.8* (n = 20)	12.3 ±0.8	6.70* ±0.04
n = 21	−66.3	3.86* ±0.08	0.84* ±0.05	3.00 ±0.10	134.1 ±5.3* (n = 20)	17.8 ±1.7* (n = 20)	13.9+ ±0.78	6.77* ±0.04
Control subjects (n = 21)	−88.8	3.45 ±0.06	0.60 ±0.02	2.86 ±0.05	154.6 ±3.9+ (n = 19)	3.8 ±0.5 (n = 19)	12.3 ±0.3	7.00 ±0.01

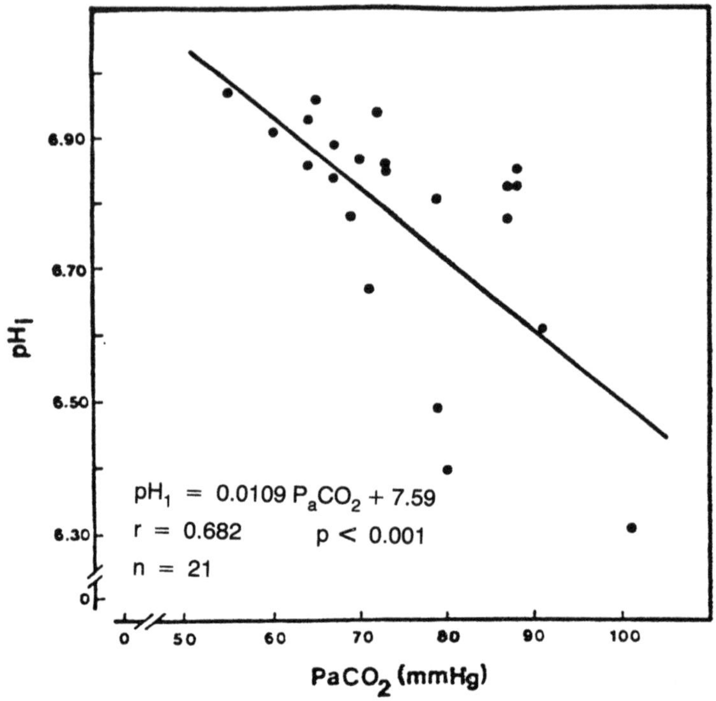

Fig. 1 - Correlation between $PaCO_2$ and intracellular pH (pHi) in COPD patients with acute respiratory failure (ARF).

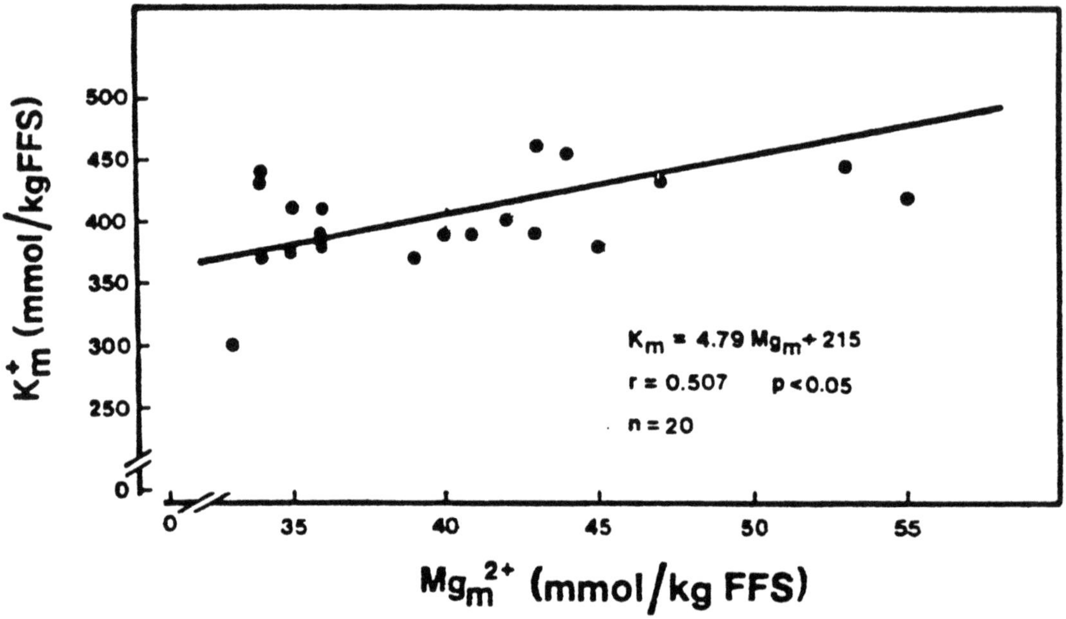

Fig. 2 - Correlation between magnesium (Mg_m^{2+}) and potassium (K_m^+) content in skeletal muscle of COPD patients with ARF.

Remarkable derangements of the energy metabolism of skeletal muscles have already been found in COPD patients, in both respiratory and non-respiratory skeletal muscles.

Should we only consider ATP and phosphocreatine values, skeletal muscle of COPD patients would appear as characterized by reduced high-energy phosphate compound content. This has been demonstrated by several authors, both in patients with COPD of moderate degree[8, 9, 10] and in patients with COPD and Acute Respiratory Failure[11, 12].

In a further study our group attempted a parallel evaluation of the energy metabolism and acid-base equilibrium in skeletal muscle of COPD patients with ARF. Ten patients with severely hypercapnic-hypoxemic COPD and ten age-matched control subjects were utilized. The experimental procedure included needle biopsies from the quadriceps femoris and the evaluation of intracellular acid-base equilibrium[5] and of the main parameters of cell energy metabolism[13].

In Table 3 energy metabolism parameters and intracellular pH values of COPD patients with ARF are shown.

A low content of high-energy phosphate compounds and a condition of intracellular acidosis are evident. In fact, a marked decrease of both muscle ATP and phosphocreatine contents is present. However, no differences from controls were found in ADP, AMP, or Total Creatine. A significant decrease of Total Adenine Nucleotide, ATP/ADP ratio and Energy Charge Potential was found. Muscle and plasma lactate were both increased, though with remarkable disparity of values. Intracellular pH values were significantly lower than control values. No significant relationship was found between energy metabolism parameters and intracellular pH or between muscle or plasma lactate, PaO_2 and intracellular pH.

At any given low PaO_2 value, a wide scattering of muscle lactate values was found (Fig. 3); in six patients muscle lactate values fell within the control range.

From both the present data on COPD patients with ARF and the previously discussed literature data on patients with COPD of various degrees (from moderate to severe), it is evident that a reduction in both ATP and PCr content characterizes the skeletal muscle of COPD patients, regardless of the type and activity levels of skeletal muscle considered or the severity of COPD.

The extent of the decrease of ATP and PCr contents (about 30-35% of control values) is comparable to that found, at rest, in patients with cardiogenic shock or severe congestive heart failure[14] and in a series of patients with severe acute respiratory or circulatory insufficiency[15].

Among the factors likely to result in an altered muscle cell energy metabolism in the COPD patients considered in our study, hypoxia and intracellular acidosis can be credited with a major role.

The COPD patients we studied were severely hypoxemic.

Low values found for ATP and phosphocreatine, as well as increased muscle and plasma lactate values, may be explained as a consequence of hypoxia in skeletal

Table 3 - Muscle energy metabolism and intracellular acid-base equilibrium in COPD patients with arf.

	PCr*	Cr*	TCr*	ATP*	ADP*	AMP*	TAN*	ATP/ADP	ECP	LACTm*	LACTpfv**	pH_i
COPD patients N. 10	49.02 ± 11.07	66.50 ± 21.06	115.60 ± 23.76	16.77 ± 2.09	3.37 ± 0.66	0.24 ± 0.12	20.61 ± 2.08	5.14 ± 1.27	0.899 ± 0.024	15.04 ± 7.5	1.44 ± 0.8	6.75 ± 0.23
Controls N. 10	76.87 ± 6.80	51.80 ± 11.05	127.68 ± 17.26	23.76 ± 1.07	2.94 ± 0.49	0.14 ± 0.08	26.84 ± 1.10	8.36 ± 1.33	0.937 ± 0.013	8.6 ± 2.6	0.81 ± 0.12	7.03 ± 0.07
Mann-Whitney U Test	0.001	0.05	ns	0.001	ns	ns	0.001	0.001	0.001	0.05	0.05	0.001

Data expressed as mean ± SD

* mmol/kg muscle dry-weight

** mmol/l of femoral vein plasma

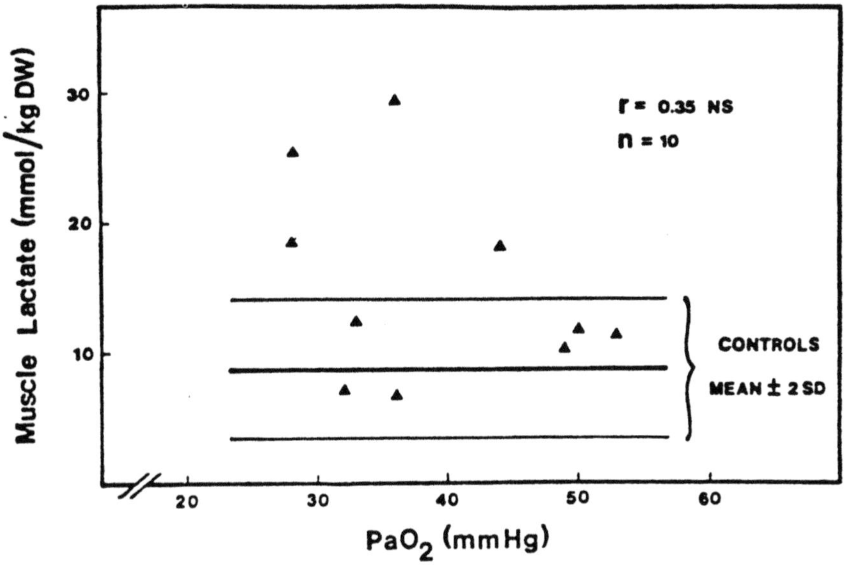

Fig. 3 - Relationships between muscle lactate content and PaO_2 values.

muscle, thus resulting in an impaired mitochondrial oxidative function.

Nonetheless, the wide scattering of muscle lactate values and the lack of correlation between muscle lactate and PaO_2 values at low PaO_2 levels point to the existence of further factors interfering with muscle metabolic response to hypoxia.

The contrasting effects of some factors possibly interfering with muscle lactate production in response to hypoxemia in our COPD patients are summarized in Figure 4.

It is well known that a reduction of arterial blood oxygen levels does not necessarily correlate with the degree of tissue hypoxia. In fact, the O_2 availability for cells is also influenced by systemic and local factors, regulating mainly O_2 exchange and microcirculation. These factors, e.g. cardiac output and its distribution or local blood supply, have not been measured in our COPD patients with ARF. Yet we cannot exclude that they may be altered in such severely ill patients.

In spite of severe hypoxemia, in several of our COPD patients muscle lactate levels were normal. But in these patients an intracellular acidosis was present. Low intracellular pH is known as determining a severe inhibition of key glycolytic enzymes phosphorilase and PFK[16].

Therefore we can hypothesize that, in spite of severe hypoxemia, in some of our COPD patients the effects of intracellular acidosis on glycolitic sequences have limited muscle lactate production.

The negative feed-back exerted by H+ on lactate production has been regarded as an important homeostatic mechanism of 'metabolic buffering'[17] and as a part

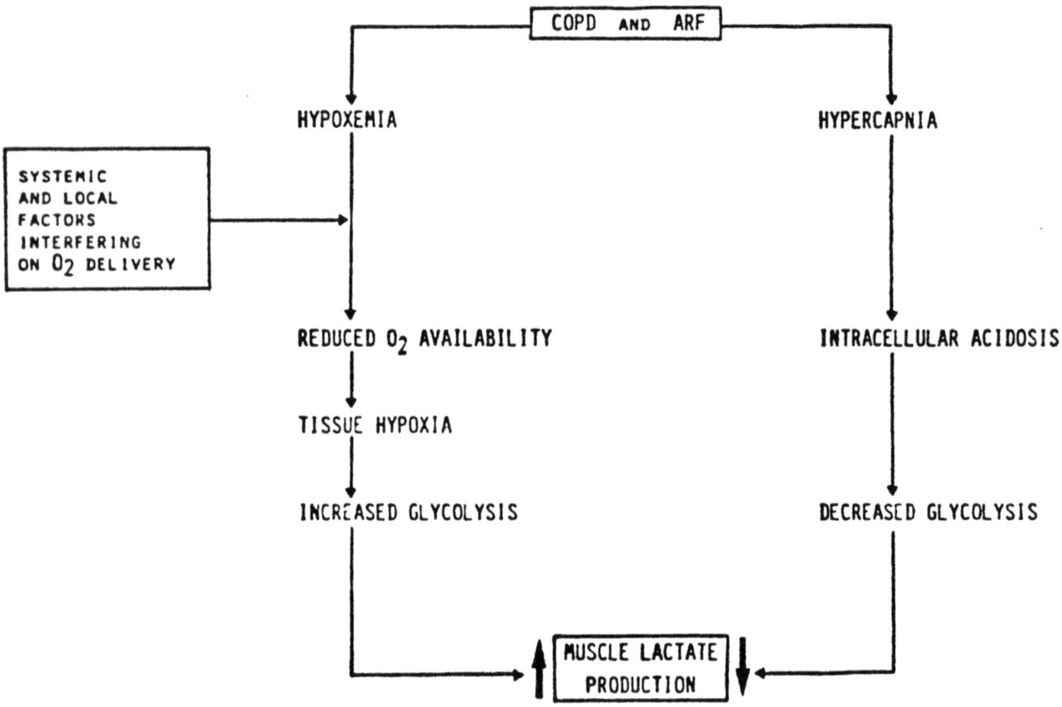

Fig. 4 - Same factors possibly interfering on muscle lactate production in COPD patients with ARF are illustrated.

of a more general homeostatic system that regulates endogenous organic acid production, via changes in systemic pH[18].

A marked suppression of exercise-induced hyperlactatemia by hypercapnia has been ascertained in man[19, 20].

The role of such negative feed-back in the pathogenesis and evolution of clinical lactic acidosis, especially in patients with both hypercapnia-related intracellular acidosis and hypoxemia, is still unclear. On the other hand, the same servomechanism is apparently overridden in cases of severely hypoxemic lactic acidosis[21], and one cannot exclude that this applies to some of our patients in whom the inhibitory effect of acidosis is overridden by the hypoxia drive to increased lactate production.

Regeneration of ATP stores could depend not only upon rephosphorilation of AMP and ADP, but also upon the availability of new adenine nucleotides. Low ATP leveis, as found in our patients, could also be ascribed to a disrupted balance between degradation and resynthesis of adenine nucleotides. In fact, both hypoxia and acidosis are known as activating ATP-degrading sequences. In particular AMP deaminase is activated by low pH values, by reduced O_2 availability and by low ATP values.[22]

Lowering of the skeletal muscle adenine nucleotide pool, along with low Phosphocreatine values, has been observed in a series of patients with malnutrition[23]. Under such conditions, the reduced content of both ATP and PCr has been attributed to lack of precursors and calories, owing to a reduced nutrient intake. Indeed, very few data are available with regard to the ATP and PCr content of tissues in malnourished patients.

Nutritional status has not been evaluated in the COPD patients considered above. Nevertheless, there is now increasing evidence that malnutrition is quite common among COPD patients, even though the prevalence and features of malnutrition in course of severely hypercapnic-hypoxemic COPD are poorly defined[24, 25].

For this reason, a cross-sectional survey on the nutritional status of COPD patients with different degrees of hypercapnia and hypoxemia was carried out by means of well-established non-invasive techniques of anthropometric and biochemical evaluating indexes.

84 patients with COPD of varying degrees were included in the study. They belonged to three different groups:

- Group A: 33 outpatients treated at a day hospital.
- Group B: 20 inpatients hospitalized for Acute Respiratory Failure which had not required mechanical ventilation.
- Group C: 31 inpatients hospitalized in an Intensive Care Unit for Acute Respiratory Failure which had required mechanical ventilation.

Nutritional status was assessed by means of an integrated approach which included anthropometric measurements and biochemical data. Nutritional anthropometric indexes, expressed as % of standard values for sex and age[26, 27], included weight, height, triceps (TSF) and subscapular skinfolds and arm muscle area (AMA). Nutritional biochemical indexes included serum albumin, transferrin and total lymphocyte count.

Clinical characteristics of the three COPD patients groups are shown in Table 4: the three groups showed blood gas values of increasing severity, in particular $PaCO_2$ values.

In Table 5 anthropometric and biochemical nutritional parameters of the three groups are presented. The two groups of COPD inpatients (B and C groups) showed no difference in anthropometric parameters. These were significantly lower than those of group A. This was especially true for % body weight and % arm muscle. In both B and C groups anthropometric parameters proved considerably reduced when compared in % terms with relevant standard values. In the outpatients group (group A) only % body weight was not different from standard values, whereas all of the other anthropometric indexes were significantly reduced (Fig. 5). A significant inverse relationship between mean $PaCO_2$ and % body weight was also found (Fig. 6).

Our study suggests that in hypercapnic-hypoxemic COPD patients the severity of nutritional alterations increases in proportion to the degree of alteration of

Table 4 - Clinical features

Patients	Group A COPD Outpatients n = 33	Group B COPD Inpatients n = 20	Group C COPD Inpatients n = 31	STATISTICS			
					Student's "t" test		
				Variance analysis	A vs B	A vs C	B vs C
Age (years)	67.3 ± 1.2	69.6 ± 1.2	69.2 ± 1.6	ns	ns	ns	ns
Disease duration years	13 ± 1.4	19 ± 2.2	26 ± 2.5	$p < 0.01$	$p < 0.05$	$p < 0.001$	$p < 0.05$
Mean $PaCO_2$ mmHg	50 ± 1.1	55 ± 1.2	61 ± 1.2	$p < 0.001$	$p < 0.05$	$p < 0.001$	$p < 0.001$
Mean PaO_2 mmHg	54 ± 1.1	53 ± 1.6	46 ± 1.4	$p < 0.01$	ns	$p < 0.001$	$p < 0.01$
Hct %	49 ± 0.7	45 ± 1.3	42 ± 1.1	$p < 0.01$	$p < 0.01$	$p < 0.01$	ns

Table 5 - Nutritional anthropometric and biochemical indices

Patients	Group A COPD Outpatients n = 33	Group B COPD Inpatients n = 20	Group C COPD Inpatients n = 31	STATISTICS			
					Student's "t" test		
				Variance analysis	A vs B	A vs C	B vs C
Body weight (% standard)	97 ± 2.4	87 ± 3.1	83 ± 3.2	$p < 0.01$	$p < 0.05$	$p < 0.001$	ns
Triceps Skinfold (% of standard)	86 ± 6.1	71 ± 5.9	66 ± 6.3	$p < 0.05$	ns	$p < 0.05$	ns
Subscapular Skinfold (% of standard)	87 ± 4.5	74 ± 5.9	61 ± 4.8	$p < 0.01$	ns	$p < 0.001$	ns
Arm muscle area (% of standard)	78 ± 3.5	60 ± 4.1	62 ± 4.7	$p < 0.05$	$p < 0.001$	$p < 0.05$	ns
Albumin g%	4.1 ± 0.07	3.4 ± 0.1	3.5 ± 0.1	$p < 0.001$	$p < 0.001$	$p < 0.001$	ns
Transferrin mg%	204 ± 6.2	184 ± 1.3	180 ± 4.3	ns	ns	ns	ns
Lymphocytes mm^3	2140 ± 101	2100 ± 100	1620 ± 75	$p < 0.001$	ns	$p < 0.001$	$p < 0.001$

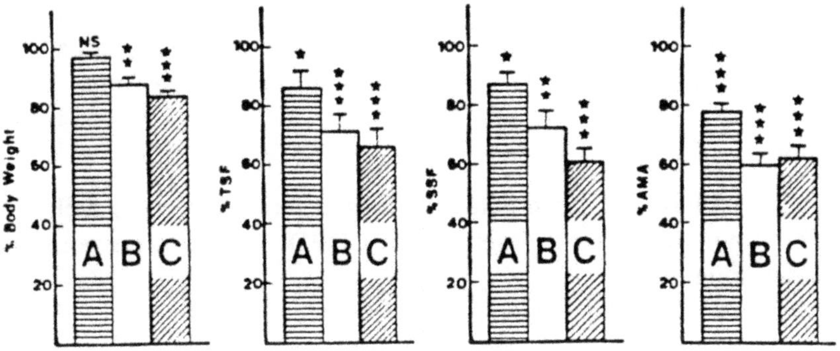

Fig. 5 - Body Weight %, Triceps Skinfold Thickness (TSF%), Subscapular Skinfold Thickness (SSF%) and Arm Muscle Area (AMA%), expressed as percent of standard values, are presented for the three groups of COPD patients
Statistical significance (single sample student's "t" test) of differences from standard values are showed (*$p < 0.05$, **$p < 0.01$, ***$p < 0.001$).

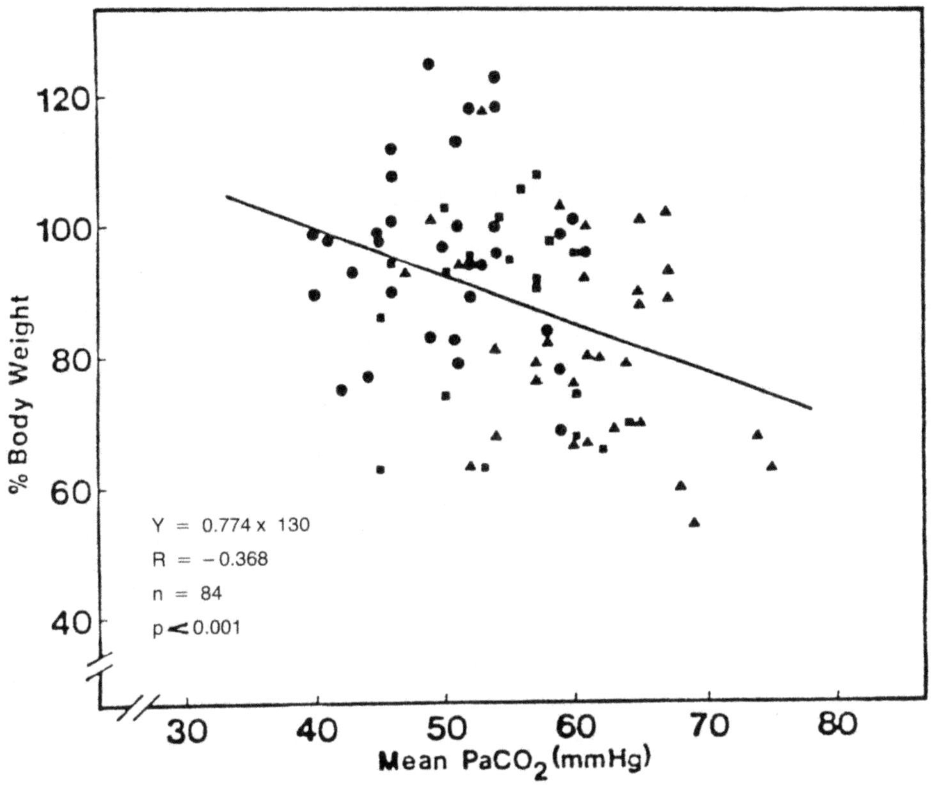

Fig. 6 - Correlation between $PaCO_2$ and % body weight.

respiratory function, as assessed from the degree of impairment of blood gas values. Depletion of both fat stores and lean body mass appears to be the hallmark of nutritional status impairment in hypercapnic-hypoxemic COPD patients.

Conclusion

To sum up, we have presented data regarding different aspects of skeletal muscle cell metabolism in the course of severe degree COPD. The following points are now worth underlining:

- The presence of intracellular acidosis, potassium depletion, low muscle magnesium content and low high energy phosphate content has been demonstrated in skeletal muscle of COPD patients with ARF considered in the present study.
- Some of the abnormalities discussed above are factors known to impair skeletal muscle function.
- In patients with COPD further investigations along the same lines should attempt to relate muscle acid-base, electrolyte and energy metabolism derangements to contractile function alterations of respiratory muscles.
- This could hopefully contribute to a better understanding of the cellular mechanisms leading to respiratory muscle fatigue and weakness.

References

1. TUSHAN F.S., BROMBERG P.A. et al: Intracellular pH and electrolyte metabolism in chronic stable stable hypercapnia. Arch. Int. Med. 1970. 125: 967-974.
2. CAMPBELL R.H.A., BRAND H.L. et al: Body weight and body water in chronic cor pulmonale. Cli. Sci. Mol. Med. 1975. 49: 323-335.
3. MOLLER P., BERGSTROM J. et al: Energy-rich phosphagens, electrolytes and free aminoacids in leg skeletal muscle of patients with chronic obstructive lung disease. Acta Med. Scand. 1982. 211: 187-193.
4. FIACCADORI E., DEL CANALE S. et al: Intracellular acid-base and electrolyte metabolism in skeletal muscle of patients with chronic obstructive lung disease and respiratory failure. Clin. Sci., 1986. 71: 6, 703-12.
5. FIACCADORI E., GUARIGLIA A., et al: Intracellular bicarbonate and pH determination in rat skeletal muscle from acid-labile CO2 measurement from dead-stop end-point potentiometric titrations. Eur. Rev. Med. Pharmacol. Sci., (Rome). 1986. 8: 3-14.
6. BERGSTROM J., ALVESTRAND A., et al: Muscle intracellular electrolytes in patients with chronic uremia. Kidney Int. 1983. 24 (S16): S153-S160.
7. CUNNINGHAM J.N. JR., CARTER N.W. et al: Resting transmembrane potential difference of skeletal muscle in normal subjects and severely ill patients. J. Clin. Invest. 1971. 50: 49-59.
8. MOLLER P., BERGSTROM J., FURST P. et al: Energy-rich phosphagens, electrolytes and free amino acids in leg skeletal muscle of patients with chronic obstructive lung disease. Acta Med. Scand. 1982. 211: 187-193.
9. CAMPBELL J.A., HUGHE R.L. et al: Alterations in intercostal muscle morphology and biochemistry in patients with obstructive lung disease. Am. Rev. Resp. Dis. 1980. 122: 679-686.
10. HUGHES R.L., KATZ H. et al: Fiber size and energy metabolites in five separate muscles from patients with COLD. Respiration 1983. 44: 321-328.

11. GERTZ I., HEDENSTIERNE G. et al: Muscle metabolism in patients with chronic obstructive lung disease and acute respiratory failure. Clin. Sci. Mol. Med. 1977. 528: 395-403.
12. DEL CANALE S., SODERLUND K., FIACCADORI E. et al: Abnormal energy metabolism in severe chronic obstructive lung disease. Eur. Rev. Respir. Dis. (Suppl.) 1986. 146, 239-43.
13. HARRIS R., HULTMAN E., et al: Glycogen, glycolytic intermediate and high energy phosphate in biopsy samples of m. quadriceps femoris of man at rest. Methods and variance of values. Scand. J. Clin. Lab. Invest. 1974. 33: 109-113.
14. KARLSSON J.T., WILLERSON S.J. et al: Skeletal muscle metabolites in patients with cardiogenic shock or severe congestive heart failure. Scand. J. Clin. Lab. Invest. 1975. 35: 73-79.
15. BERGSTORM J., BOSTROM H. et al: Preliminary studies of energy-rich phosphagens in muscle from severely ill patients. Crit. Care Med. 1976. 4: 197-204.
16. RELMAN A.S.: Metabolic consequences of acid-base disorders. Kidney Int. 1972. 1: 347-357.
17. COHEN R.D., ILES R.A.: Intracellular pH: measurement, control and metabolic interrelationships. CRC Crit. Rev. Clin. Lab. Sci. 1975. 6: 101-142.
18. HOOD V.L., TANNER R.L.: pH control of lactic and keto acid production: a mechanism of acid-base regulation. Min. Electr. Metab. 1983. 9: 317-325.
19. GRAHAM T.E., WILSON B.A. et al: The effects of hypercapnia on the metabolic response to steady-state exercise. Med. Sci. Sports Exerc. 1982. 14: 286-291.
20. EHRSAM R.E., HEIGHENHAUSER G.J. F. et al: Effect of respiratory acidosis on metabolism in exercise. J. Appl. Physiol. 1982. 53 (1): 63-69.
21. MADIAS N.E.; Lactic acidosis. Kidney Int. 1986. 29: 752-774.
22. FOX I.H.: Metabolic basis for disorders or purine nucleotide degradation. Metabolism. 1981. 30 (6): 616-634.
23. FURST P., BERGSTROM J. et al: Intermediate energy metabolism for the catabolic state with special regard to muscle tissue. In: Wilkinson A.W., Cuthbertson D. (Eds): Metabolism and the response to injury. London Pitman Medical, p. 94. 1976.
24. DRIVER G.A., LE BRUN M.: Iatrogenic malnutrition in patients receiving ventilatory support JAMA 244: 1980. 2195-2196.
25. DRIVER A.G., MC ALEVY M.T. et al: Nutritional assessment of patients with chronic obstructive pulmonary disease and acute respiratory failure. Chest 1982. 82: 568-571.
26. FRISANCHO A.R.: New standards of weight and body composition by frame size and height for assessment of nutritional status of adults and the elderly. Am. J. Clin. Nutr. 1984. 40: 808-819.
27. HEYMSFIELD SB, MC MANUS C. et al: Anthropometric measurement of muscle mass: revised equations for calculating bone-free arm muscle area. Am. J. Clin. Nutr. 1982. 36: 680-690.

A. GRASSINO - C. FRACCHIA - C. RAMPULLA - L. ZOCCHI
Respiratory Muscles in COPD
Bi & Gi Publishers, Verona - Springer-Verlag, London, 1988

Inspiratory muscles and dyspnea

J.W. FITTING

Division de pneumologie, Départment de médecine interne, Centre Hospitalier Universitaire Vaudois, Lausanne, Switzerland

SUMMARY: The relationship between dyspnea, defined as a sensation of laboured or difficult breathing, and respiratory muscle activity is discussed.
While the role of rib cage muscle fatigue in dyspnea has yet to be fully studied, diaphragmatic fatigue related to the diaphragm's status as an inspiratory muscle is probably a contributory factor.
Fatigue of the diaphragm in its function as an abdominal explusive muscle will, by contrast, not affect perception of breathing load.
The lungs and upper airways are not normally thought to contribute directly to respiratory sensation, though data relating to the latter are conflicting.
Increased ventilatory impedance, as in COPD, will, of course, necessitate increased respiratory effort, with attendant inspiratory muscle fatigue and sensation of laboured breathing.

KEY WORDS: Inspiratory muscle; dyspnea; breathing; respiratory muscle activity; diaphragmatic fatigue; abdominal expulsive muscle; breathing load; ventilatory impedance, increased; COPD; respiratory effort; inspiratory muscle fatigue; laboured breathing.

Breathing is a rhythmic and automatic activity which occurs without conscious awareness in normal subjects, at least at rest. In some circumstances, this activity can be sensed and it is then referred to as dyspnea, felt as a sensation of laboured or difficult breathing[2]. According to this broad definition, dyspnea can occur in very different conditions such as high-level exercise or loaded breathing in normal subjects, minor exercise or rest in patients with COPD, pulmonary fibrosis, heart failure, anemia, neuro-muscular disorders, etc. Thus, if a single sensory mechanism is to be found, it should account for dyspnea in all these situations. The technique of loaded breathing has been most frequently used in studies on respiratory sensations, since it enables the establishment of a relationship between a quantified stimulus and the sensation it evokes.

The major potential origins which have been considered for respiratory sensations are the lungs, the upper airways, the rib cage, the respiratory muscles and chemical drive. The lungs are usually denied a role in respiratory sensation, since bilateral vagal block has been shown to alter neither the threshold of load detection nor the sensation associated with higher external loads[18]. The data pertaining to the upper airways are more conflicting. On the one hand, it was observed in one patient that load detection was impaired during breathing through a tracheostomy tube, whereas it was normal when the intraluminal pressure was allowed to be transmitted to the upper airways[26]. On the other hand, load detection has been found not to be modified after local anesthesia of the upper airways[7, 8]. This may be explained by the observation that the threshold for pressure detection at the mouth is higher when the pressure change is confined to the upper airways and, thus, not transmitted to the thoracic structures[13]. It should be noted that these studies used load detection and, therefore, refer to very low pressures. The situation may be different with high pressures. However, the upper airways probably play no role in usual dyspnea, since only external loading creates large subatmospheric pressures at this level. Threshold studies also gave conflicting results about the sensory role of the rib cage, load detection being either altered[25] or not[27] in patients with upper spinal cord lesion and not altered by spinal anesthesia[9]. The observation by Gottfried et al.[14] that load perception of quadriplegic patients is impaired when studied over a wider range of loads supports the hypothesis that afferents from the rib cage receptors play a sensory role.

Two studies clearly demonstrate that the respiratory muscles play an important role in respiratory sensations. In the first, resistive load detection was estimated in normal subjects during spontaneous breathing and during passive ventilation by an external negative-pressure respirator. The detection threshold was markedly elevated in the latter circumstance, indicating that load detection depends on active respiratory muscle contraction[18]. In the second study, Campbell et al.[5] found that the size of resistive loads was overestimated when respiratory muscles were weakened by partial curarization.

The question as to whether hypercapnic drive plays a sensory role in itself has been recently investigated by Castele et al.[6]. Normal subjects were hyperventilated with a positive-pressure respirator with fixed values of tidal volume and frequency. The PCO_2 was progressively increased by raising the inspired CO_2 concentration and the subjects were asked to signal when they first sensed a need for increased ventilation. It was observed that the onset of inspiratory muscle activity always preceded the awareness of ventilatory need, suggesting that the sensation evoked by increased ventilatory drive is related to muscle activity and not to chemical drive *per se*.

Using psychophysical techniques, Altose et al.[1] and Killian et al.[19] showed that the relationship between the inspiratory load and the evoked sensation follows a power law as in other sensory modes. Killian et al.[20] demonstrated, then, that the perceived magnitude of added loads is not a function of the load itself but is directly related to the inspiratory pressure, its duration and probably, therefore, to inspiratory muscle force and its duration.

If the role played by inspiratory muscles in respiratory sensation has been established, the question of the receptors and neural pathways involved remains open. In limb muscles, there appears to exist both a sense of force and a sense of effort.

The sense of force would originate from muscle receptors, tendon organs and perhaps spindles, and the sense of effort from corollary discharges of efferent motor command. However, the exact pathways are still unknown and are probably more complex, since the perception of effort appears to need some peripheral feedback and the perception of force to require some central information for interpretation of spindle afferent discharges[24]. Effort and force normally increase in parallel, but effort increases disproportionately with force when a muscle is weakened, as by operating at a suboptimal length[4] or during partial neuro-muscular blockade[12] or fatigue[23].

Dyspnea occurs in three major circumstances: increased ventilation (e.g. exercise), increased respiratory impedance (e.g. pulmonary diseases) and muscule weakness (e.g. neuro-muscular disorders). The sense of respiratory force would account for dyspnea in the first two conditions but not in the last one, where respiratory pressures are low. On the contrary, the sense of respiratory effort is increased in all three circumstances and is, therefore, theoretically more likely to promote the sensation of dyspnea[22].

Some experimental evidence supports this hypothesis. Firstly, large interindividual differences in dyspnea exist at similar levels of inspiratory pressure during resistive loading. For a given pressure, dyspnea is less for a subject with strong respiratory muscles than for another with weaker muscles. This variability is reduced if dyspnea is related to pressure expressed as a fraction of maximal inspiratory pressure[17]. This observation is better explained if dyspnea is determined by respiratory effort than if it is related to respiratory force. Secondly, Killian et al.[21] reported that subjects are able to distinguish between the sensations of tension, effort and breathlessness during elastic loading at different lung volumes. The relationship between the sensation of tension and inspiratory pressure was similar at FRC and at higher lung volume. On the other hand, the sensations of effort and breathlessness, at a given pressure, increased at higher volumes, presumably because inspiratory muscle strength was reduced by a shorter length. Recently, the same group reported that the perceived magnitude of respiratory effort increases for a given pressure if inspiratory flow (reflecting velocity of shortening) and tidal volume (reflecting extent of shortening) increase[10].

If dyspnea is related to the sense of respiratory effort, it can be expected to increase with respiratory muscle fatigue as manifested by reduced respiratory muscle strength. Studies of the relationship of dyspnea to respiratory muscle fatigue are scarce and, at first sight, contradictory. Supinski et al.[28] reported that respiratory effort increases with time as electromyographic signs of fatigue develop in inspiratory muscles during inspiratory loading. On the other hand, Grassino et al.[15] observed that marked dyspnea could be induced by exercise in patients with COPD without any signs of fatigue of the diaphragm. Conversely, electromyographic signs of diaphragmatic fatigue developed without marked dyspnea when the same patients followed an imposed breathing pattern with increased tidal volume and inspiratory

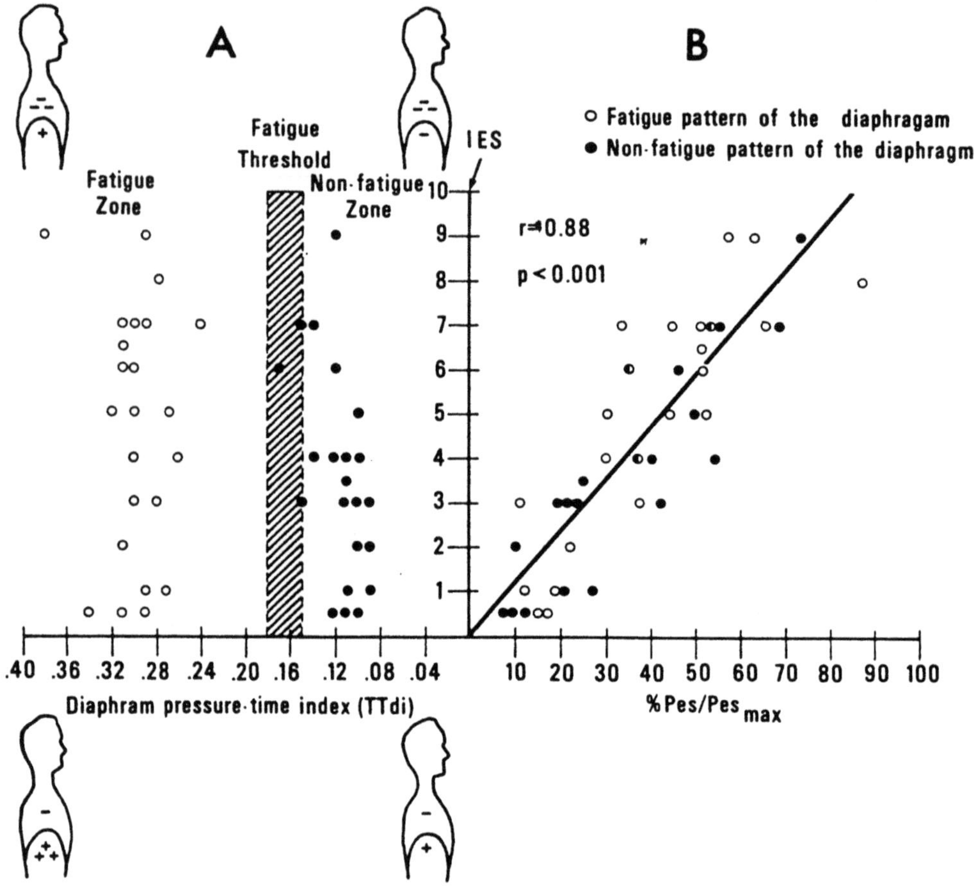

Fig. 1. - Effect of fatiguing and non-fatiguing patterns of the diaphragm on inspiratory effort sensation. A: relationship between insipiratory effort sensation (IES) and the tension-time index of the diaphragm (TTdi = Pdi/PdiMAX · T_T/T_{TOT}. TTdi values above 0.15-0.18 correspond to fatiguing patterns of the diaphragm.
B: relationship between IES and esophageal pressure swings expressed as percentage of maximal inspiratory pressure. [3]

time. To solve this question, Bradley et al.[3] formally studied the influence of fatigue development of the diaphragm on the sensation of inspiratory effort. While breathing against different inspiratory resistances, normal subjects were asked to perform either fatiguing or non-fatiguing diaphragmatic patterns with various combinations of pleural (Ppl) and transdiaphragmatic (Pdi) pressures. The sensation of inspiratory effort was found to be related to Ppl and to be independent of the presence of a fatiguing diaphragmatic pattern (Fig. 1). In other words, high inspiratory effort sensation could occur with non-fatiguing contractions of the diaphragm and low inspiratory effort sensation with fatiguing contractions of the diaphragm. This does not imply that diaphragmatic contractions are not sensed. When breathing with high Pdi and low Ppl swings, the subjects reported an uncomfortable feeling in the epigastric area which was not sensed as a difficulty on inspiration.

Fitting et al.[11] studied the contribution of various muscles to inspiratory effort sensation by asking subjects to emphasize the use of the diaphragm or the rib cage muscles or a combination of both against inspiratory resistances. They found a unique relationship between inspiratory effort sensation and Ppl, regardless of the muscle group contributing to this pressure. Thus, under these experimental conditions, diaphragmatic contractions elicited a sensation of inspiratory effort. These two studies suggest that diaphragmatic contractions must give rise to increased pleural pressure swings to be interpreted as an inspiratory effort. This is not in contradiction with the aforementioned theories, since some peripheral information appears to be necessary for the efferent motor command to be sensed as an effort[24]. In this case, the afferent feed-back information signalling increased Ppl swings would originate from some extra-diaphragmatic structure in the thorax. This mechanism would account for the double function of the diaphragm, which can act on the thorax as an inspiratory muscle or on the abdomen as an expulsive muscle. If the receptors supplying the afferent feed-back signals are not located in the diaphragm itself but in the thorax and in the abdomen, appropriate interpretation of diaphragmatic efforts is possible in each of its functions.

According to this theory, the effect of diaphragmatic fatigue on dyspnea will depend on which function the diaphragm is fulfilling. If fatigue develops while the diaphragm is acting as an inspiratory muscle, it may contribute to dyspnea. If fatigue develops while the diaphragm is acting as an abdominal expulsive muscle, it will not be sensed as dyspnea. The role of rib cage muscle fatigue in the genesis of dyspnea remains to be studied.

In conclusion, dyspnea appears to be promoted by the activity of inspiratory muscles and, more precisely, to correspond to the sense of inspiratory effort. COPD patients breathe with an abnormally high respiratory effort for several reasons. Their respiratory impedance is increased because of elevated airway resistance and reduced dynamic lung compliance. Their minute ventilation may be increased because of an increased dead space ventilation. Finally, the capacity of their respiratory muscles is reduced since they operate at suboptimal length and configuration because of hyperinflation. All these factors may lead to fatigue, which further weakens the respiratory muscles and requires a greater increase of respiratory effort.

References

1. ALTOSE M.D., CHERNIACK N.S., Respiratory sensation and respiratory muscle activity. In: Hutas I., Debreczeni L.A. (eds) Adv. Physiol. Sci. Vol. 10. Respiration 111-119, 1981.

2. ALTOSE M.D.: Assessment and management of breathlessness. Chest 1985. 88: 77S-82S.
3. BRADLEY T.D., CHARTRAND D.A., FITTING J.W., KILLIAN K.J., GRASSINO A.: The relation of inspiratory effort sensation to fatiguing patterns of the diaphragm. Am. Rev. Respir. Dis.: In press.
4. CAFARELLI E., BIGLAND-RITCHIE B.: Sensation of static force in muscles of different length. Exp. Neurol. 1979. 65: 511-525.
5. CAMPBELL E.J.M., GANDEVIA S.C., KILLIAN K.J., MAHUTTE C.K., RIGG J.R.A.. Changes in the perception of inspiratory resistive loads during partial curarization. J. Physiol. (London) 1980. 309: 93-100.
8. CASTELE R.J., CONNORS A.F., ALTOSE M.D.: Effects of changes in CO2 partial pressure on the sensation of respiratory drive. J. Appl. Physiol. 1985. 59: 1747-1751.
7. CHAUDHARY B.A., BURKI N.K.: Effects of airway anesthesia on the ability to detect added inspiratory resistive loads. Clin. Sci. 1978. 54: 621-626.
8. CHAUDHARY B.A., BURKI N.K.: The effects of airway anesthesia on detection of added inspiratory elastic loads. Am. Rev. Respir. Dis. 1980. 122: 635-639.
9. EISELE J., TRENCHARD D., BURKI N., GUZ A.: The effect of chest wall block on respiratory sensation and control in man. Clin. Sci. 1968. 35: 23-33.
10. EL-MANSHAWI A., SUMMERS E., CAMPBELL E.J.M., KILLIAN K.J.. Effect of velocity and extent of shortening on the perceived magnitude of respiratory muscle effort (abstract). Am. Rev. Respir. Dis. 1986. 133: A189.
11. FITTING J.W., CHARTRAND D.A., BRADLEY T.D., KILLIAN K.J., GRASSINO A.: Effect of thoraco-abdominal breathing patterns on inspiratory effort sensation. J. Appl. Physiol.: In press.
12. GANDEVIA, S.C., MCCLOSKEY D.I.: Changes in motor commands, as shown by changes in perceived heaviness, during partial curarization and peripheral anesthesia in man. J. Physiol. (London). 1977. 272: 673-689.
13. GANDEVIA S.C., KILLIAN K.J., CAMPBELL E.J.M.: The contribution of upper airway and inspiratory muscle mechanisms to the detection of pressure changes at the mouth in normal subjects. Clin. Sci. 1981. 60: 513-518.
14. GOTTFRIED S.B., LEECH I., DIMARCO A.F., ZACCARDELLI W., ALTOSE M.D.: Sensation of respiratory force following low cervical spinal cord transection. J. Appl. Physiol. 1984. 57- 989-994.
15. GRASSINO A., BELLEMARE F., LAPORTA D.: Diaphragm fatigue and the strategy of breathing in COPD. Chest. 1984. 85: 51S-54S.
16. GUZ A., NOBLE M.I.M., WIDDICOMBE J.G., TRENCHARD D., MUSHIN W.W., MAKE A.R.: The role of vagal and glossopharyngeal afferent nerves in respiratory sensation, control of breathing and arterial pressure regulation in conscious man. Clin. Sci. 1966. 30: 161-170.
17. JONES G.L., KILLIAN K.J., SUMMERS E., JONES N.L.: Inspiratory muscle forces and endurance in maximum resistive loading. J. Appl. Physiol. 1985. 58: 1608-1615.
18. KILLIAN K.J., MAHUTTE C.K., CAMPBELL E.J.M.: Resistive load detection during passive ventilation. Clin. Sci. 1980. 59: 493-495.
19. KILLIAN K.J., MAHUTTE C.K., CAMPBELL E.J.M.: Magnitude scaling of externally added loads to breathing. Am. Rev. Respir. Dis. 1981. 123: 12-15.
20. KILLIAN K.J., BUCENS D.D., CAMPBELL E.J.M.: Effect of breathing patterns on the perceived magnitude of added loads to breathing. J. Appl. Physiol. 1982. 52: 578-584.
21. KILLIAN K.J., GANDEVIA S.C., SUMMERS E., CAMPBELL E.J.M.: Effect of increased lung volume on perception of breathlessness, effort, and tension. J. Appl. Physiol. 1984. 57: 686-691.
22. KILLIAN K.J., CAMPBELL E.J.M.: Dyspnea. In: ROUSSOS CH., MACKLEM P.T., (Eds.). Lung Biology in Health and Disease, New York, Dekker, 29: 787-828, 1985.
23. MCCLOSKEY D.I., EBELING P., GOODWIN G.M.: Estimation of weights and tension and apparent involvement of a 'sense of effort'. Exp. Neurol. 1974. 42: 220-232.
24. MCCLOSKEY D.I.: Kinesthetic sensibility. Physiol. Reviews. 1978. 58: 763-820.
25. NEWSOM DAVIS J.: Contribution of somatic receptors in the chest wall to detection of added inspiratory airway resistance. Clin. Sci. 1967. 33: 249-260.

26. NOBLE M.I.M., EISELE J.H., TRENCHARD D., GUZ A.: Effect of selective peripheral nerve blocks on respiratory sensation. In: Breathing: Hering-Breuer Centenary Symposium, Ciba Foundation Symposium, 233-251, 1970.
27. NOBLE, M.I.M., FRANKEL H.L., ELSE W., GUZ A.: The ability of man to detect added resistive loads to breathing. Clin. Sci. 1971. 41: 285-287.
28. SUPINSKI G.S., CLARY S., BARK H., ALTOSE M., KELSEN S.G.: Effect of inspiratory muscle fatigue on the sense of effort during loaded breathing (abstract). Fed. Proc. 1985. 44: 1352.

A. GRASSINO - C. FRACCHIA - C. RAMPULLA - L. ZOCCHI
Respiratory Muscles in COPD
Bi & Gi Publishers, Verona - Springer-Verlag, London, 1988

Respiratory muscle weakness

M. GREEN, A.K. MIER, J. MOXHAM
Brompton Hospital, London, U.K.

SUMMARY: Respiratory muscle weakness, while not always subject to ready clinical identification, is symptomatic of a broad range of muscular and nervous disorders.
It can be quantitatively tested by evaluation of transdiaphragmatic pressure during sniffs or during phrenic nerve twitches.
Such testing obviously has important implications, both diagnostically and in terms of providing suitable therapy.

KEY WORDS: Weakness, respiratory muscle; nervous disorders; muscular disorders; transdiaphragmatic pressure; sniffs; phrenic nerve twitches; diaphragm weakness; myasthenia gravis.

The respiratory muscles provide the motive power for ventilation. They are skeletal muscles whose biochemistry and contractile physiology is similar to that of limb muscles. It is well known that limb muscles can become weak in a wide variety of disorders. Myopathies can effect the muscles themselves, myasthenia gravis affects the neuromuscular junction, neuropathies impair the function of peripheral nerves and motor neurone disease affects the anterior horn cells. Disorders of the central nervous system can affect the spinal cord, brain stem, and cerebral cortex, as well as the co-ordinating mechanisms in the cerebellum. All of these can cause limb muscle weakness. Finally, systemic disorders such as malnutrition or malabsorption, electrolyte imbalances, endocrine disorders and drugs can affect limb muscle performance.

It seems likely that the respiratory muscles with their similar contractile physiology might also be impaired in the types of disorder outlined above. Whilst complete diaphragm and respiratory muscle paralysis has been long recognised[7], lesser degrees of respiratory muscle weakness have not been widely investigated, nor are they often recognised clinically[3]. However, such weakness may have important consequences for ventilation and may contribute to breathlessness and respiratory insufficiency, both in patients with normal lungs and, more particularly, in those with concomitant pulmonary dysfunctions.

The strength of a limb muscle such as the quadriceps can be measured by seating the subject in a purpose-designed chair and measuring the force produced by maximum voluntary contraction of the thigh muscles[2]. By this means a range of normal values of quadriceps strength has been built up, with a well established lower limit. Thus it is possible to identify weakness of the quadriceps, and to quantitate this in terms of absolute force generated between complete paralysis and normal force. We have been attempting recently to quantity respiratory muscle function similarly.

Detection of respiratory muscle weakness

Clinically, the first step in detecting respiratory muscle weakness is to think of this possibility. There may be some pointers from the patients' history, such as breathlessness, impaired cough or unexplained respiratory infections, although these are relatively non-specific symptoms. Occasionally a patient may complain of muscle weakness elsewhere or of orthopnoea. These should alert the clinician. The patient should be carefully examined for general neuromuscular abnormalities. In the respiratory system there is sometimes obvious wasting of intercostal and accessory muscles. There may be abnormalities of chest wall movement and in particular paradoxical movement of the abdomen, that is, inward motion of the abdominal wall during quiet relaxed inspiration when supine.

Investigations

If weakness is suspected it is important to investigate the patient in more detail, and we have been analysing which tests are most helpful.

The plain chest x-ray may show raised diaphragms and small lung fields when there is diaphragm weakness, but often appearances are essentially normal. Even screening the diaphragm, which must be carried out supine, may show apparently normal movements, and it is hard to quantity reduction of movement.

Reduction of vital capacity frequently occurs when there is muscle weakness, but this may not be striking. With severe diaphragm weakness there is likely to be a fall in vital capacity from the upright to supine posture. In a recent study[9] we have looked at the fall in vital capacity between upright and supine postures (FVC, Fig. 1). Delta FVC can be up to about 20% in normal patients and patients with restrictive lung disease, and up to 30% in patients with chronic airways obstruction but without apparent muscle weakness. Here again, mild degrees of diaphragm weakness would be difficult to identify.

Global function of the respiratory muscles is reflected in the maximum inspiratory (PImax) and expiratory (PEmax) mouth pressures[1]. These manoeuvres are not difficult for co-operative subjects, but the normal values are wide, with a lower limit or PImax of as little as 40 cm H_2O in males and 30 cm H_2O in females[8]. Patients can have difficulties in carrying out the manoeuvres, particularly if there is co-existing weakness of the mouth or cheek muscles. A strong PImax, say 80 cm H_2O, makes it unlikely that there is important inspiratory muscle weakness, but lower levels are

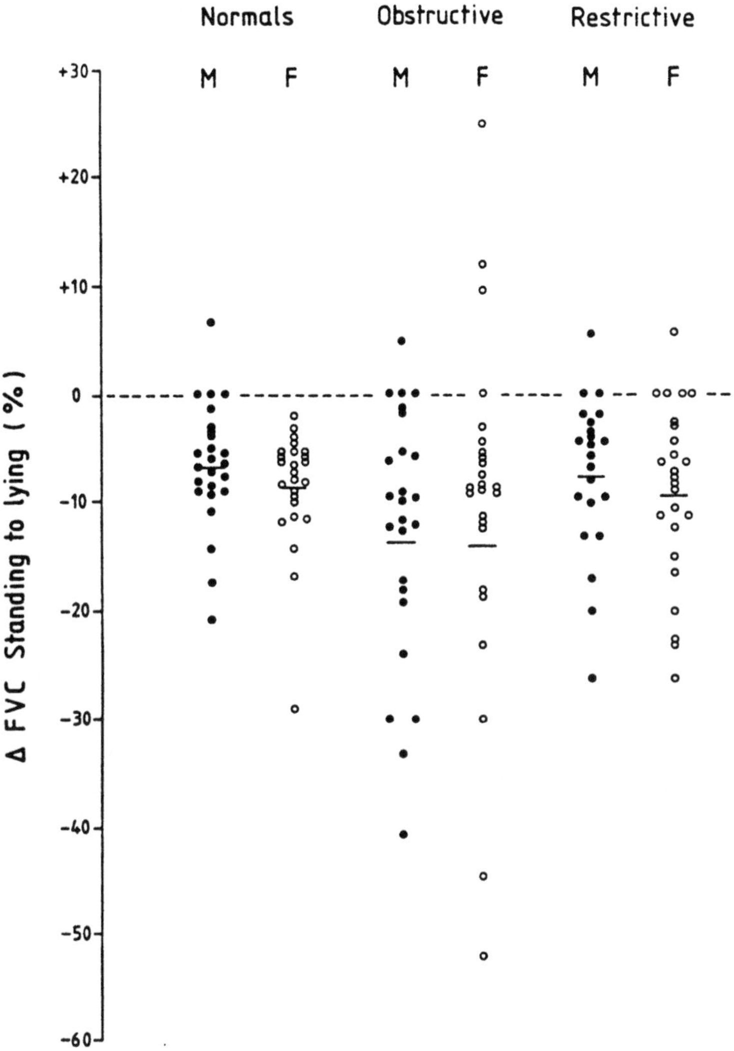

Fig. 1. - Change in FVC from standing to lying in 50 normal subjects, 50 patients with chronic airways obstruction and 47 patients with restrictive lung disease.

compatible both with normality and with moderate weakness.

Direct measurements of the forces produced by the intercostal muscles are not practical. Thus most attempts to quantify respiratory muscle strength have focused on the diaphragm. This is the main muscle of inspiration, and the pressure generated across it can be measured from gastric (Pg) and oesophageal pressures (Poes) assessed with balloon catheters (Fig. 2).

Transdiaphragmatic pressure (Pdi) is calculated as Pg - Poes with zero Pdi at resting end-expiration. This pressure has been shown in dogs to reflect closely the tension produced in the contracting muscle[4]. We have been investigating the various

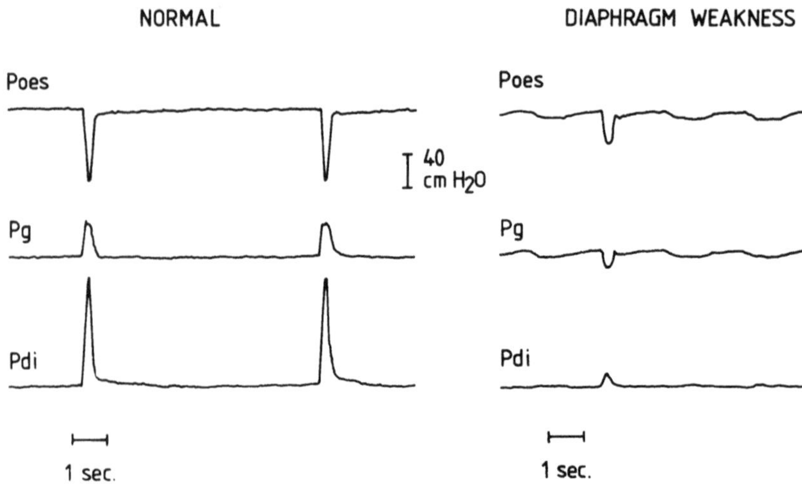

Fig. 2. - Left shows oesophageal pressure (Poes), gastric pressure (Pg) and transdiaphragmatic pressure (Pdi) during a normal sniff. On the right is the sniff in a patient with diaphragm weakness caused by trauma to the phrenic nerves at surgery.

manoeuvres for producing a maximal Pdi. We have found that Pdi during a maximum inspiration to TLC or a maximum inspiratory effort from residual volume or FRC have disadvantages. These manoeuvres are not always easy for patients, have considerable variability of Pdi and relatively low lower limits of normal. We have found that a maximum voluntary unencumbered sniff appears to be the most satisfactory manoeuvre for testing diaphragm function (Fig. 2)[5]. Whilst the normal range is relatively large, the lower limit of normal in males is 98 cm H_2O and in females is 78 cm H_2O. Figure 2 also shows the sniff in a patient with diaphragm weakness due to trauma to the phrenic nerve at surgery. Pdi is low and there is negative gastric pressure on sniffing, compared with the normal positive deflection. This man's sniffs increased in strength over the subsequent 20 months, as his phrenic nerve function returned. His symptoms improved concomitantly. Figure 3 shows sniff Pdi in a patient with myasthenia gravis. Her pre-tensilon sniffs are low, at 45 cm H_2O, and indeed she complained of breathlessness. Tensilon increased her sniffs to 65 cmH_2O, with a return to control values when the tensilon had worn off. In a recent study of 12 women with myasthenia gravis, we have found sniff Pdi to be reduced in 5, and to increase significantly with tensilon.

In order to correlate clinical features and investigations we have recently analysed 30 patients with breathlessness who were referred with suspected diaphragm weakness and were found to have sniff Pdi below normal, confirming diaphragm dysfunction. These patients had a wide variety of diagnoses including myopathies, myasthenia, neuropathies and polio. The 12 weakest patients, with sniff Pdi of 30 cm H_2O or less, had orthopnoea and abdominal paradox. This is compatible with an inability of the diaphragm to generate sufficient force to prevent the abdominal

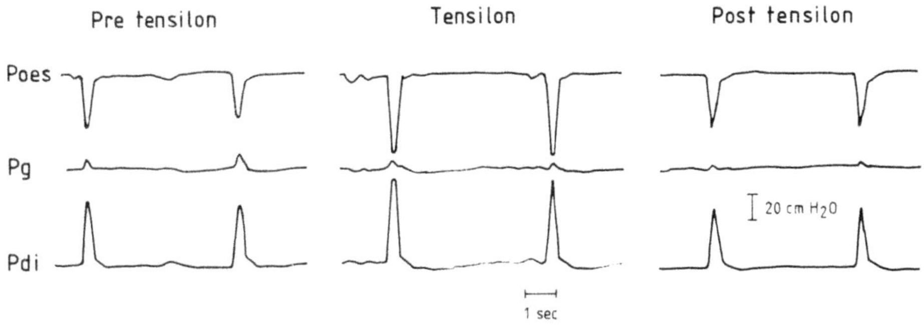

Fig. 3. - Sniffs in a 43 year old lady with myasthenia gravis. Abbreviations as Fig. 3.

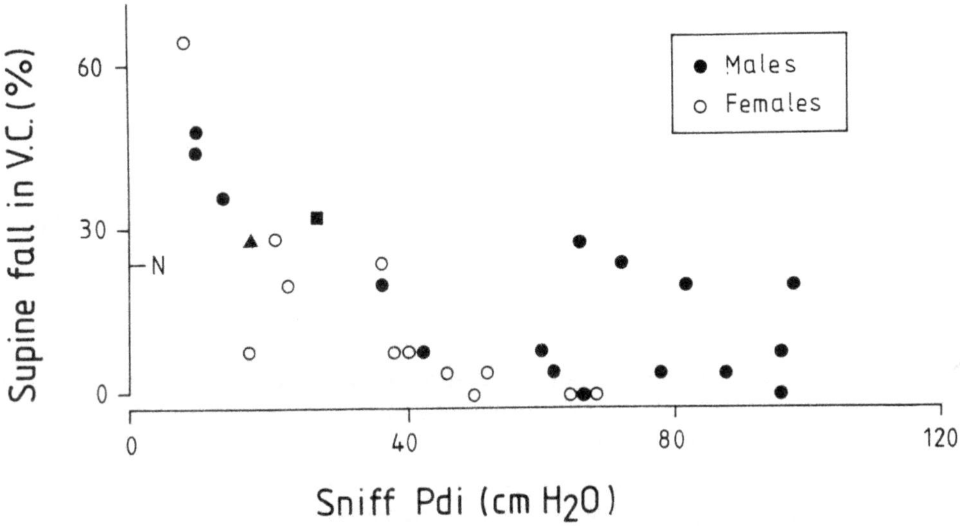

Fig. 4. - Relationship between fall in VC from standing to supine with sniff Pdi in 30 patients with diaphragm weakness.

contents being sucked upwards during inspiration when supine. However, all but 2 of the patients with less severe diaphragm weakness had neither paradox nor orthopnoea. Presumably their diaphragms, although weak, were able to counteract the weight of the abdominal contents. There was only a poor correlation between seated vital capacity and their degree of weakness, and no correlation between the level of breathlessness and their weakness as assessed by the sniff Pdi. This confirms that, whilst reduction in vital capacity and breathlessness may suggest diaphragm weakness, they are in quantitative terms poorly correlated with the level of diaphragm dysfunction. The supine fall in VC showed a better correlation (R = 0.7), (Fig. 4) but even this does not discriminate normality from weakness reliably. More specific tests, such as the measurement of sniff Pdi, need to be carried out

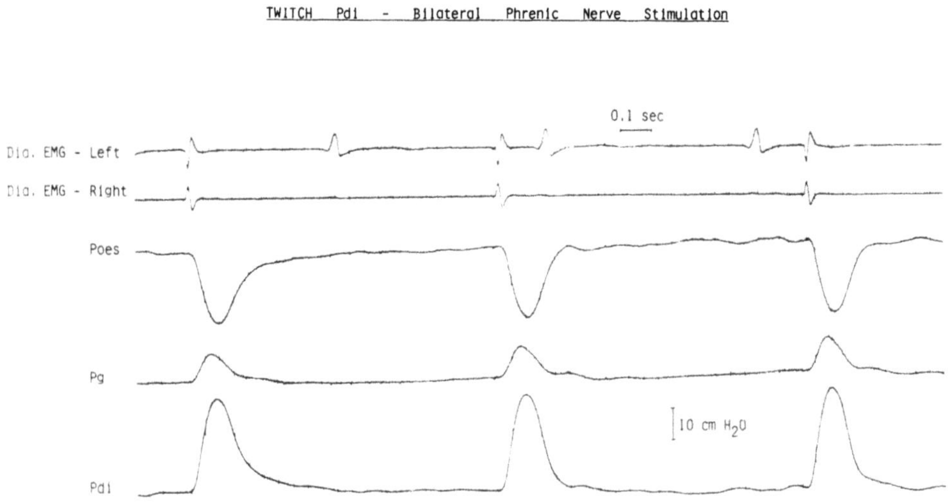

Fig. 5. - Normal diaphragm twitches generated by phrenic nerve stimulation in the neck. Abbreviations as Fig. 3.

in any patient in whom the diagnosis of diaphragm or respiratory muscle weakness is suspected.

Diaphragm twitches

Transdiaphragmatic pressure can be measured whilst stimulating the phrenic nerve electrically in the neck. We used transcutaneous stimulation with surface electrodes at 1 Hz to generate twitches and we measured the twitch Pdi (Fig. 5). The Pdi during bilateral stimulation correlated reasonably with sniff Pdi ($R = 0.75$).

Bilateral twitch Pdi averaged 18% of the maximal sniff Pdi, which is close to the twitch: tetanus ratio reported for isolated diaphragm muscle strip. This is compatible with most, if not all, phrenic nerve fibres being excited during transcutaneous phrenic nerve stimulation. The technique allows confirmation of patient effort during voluntary sniffs, since occasionally patients present with 'weakness' of hysterical origin. Stimulation of the phrenic nerve in the neck allows phrenic nerve conduction time to be measured[6], and we find this to have an upper limit of 9.5 msec in normals. If the conduction time is prolonged it suggests phrenic nerve neuropathy.

Conclusions

Respiratory muscle weakness can be hard to identify clinically. However, it appears to be common in a wide variety of systemic disorders and conditions affecting muscles, as well as the central and peripheral nervous systems. Quantification requires specific testing, and will usually include measurement of transdiaphragmatic pressure during sniffs, and possibily during phrenic nerve twitches. It is only by such quantification that it will be possible to clarify the role of the respiratory muscles and to assess their contribution to patients with respiratory dysfunction. These investigations not only have important diagnostic consequences, but may in due course lead to significant therapeutic regimes.

References

1. BLACK L.F., HYATT R.E.: Maximal respiratory pressures: normal values and relationship to age and sex. Am. Rev. Respir. Dis., 1969. 99: 696-702.
2. EDWARDS R.H.T., YOUNG A., HOSKING G.P., JONES D.A.: Human skeletal muscle function: description of tests and normal values. Clin. Sci. Mol. Med. 1977. 52: 283-290.
3. GREEN M., MOXHAM J.: The respiratory muscles. In: Flenley D.C., Petty T.L. (Eds.). Recent advances in respiratory medicine. Edinburgh. Churchill-Livingstone, 1983. 1-20.
4. KIM M.J., DRUZ W.S., DANON J. MACHNACH W., SHARP J.T.: Mechanics of the canine diaphragm. J. Appl. Physiol. 1986. 41: 369-382.
5. MILLER J., MOXHAM J., GREEN M.: The maximal sniff in the assessment of diaphragm function in man. Clin. Sci. 1985. 69: 91-96.
6. NEWSOM-DAVIS J.: Phrenic nerve conduction in man. J. Appl. Physiol. 1967. 30: 420-426.
7. NEWSOM-DAVIS J., GOLDMAN M., LOH L., CASSON M.: Diaphragm function and alveolar hypoventilation. Q. J. Med. 1976. 45: 87-100.
8. WILSON S.H., COOKER N.T., EDWARDS R.H.T., SPIRO S.G.: Predicted normal values for maximal respiratory pressures in Caucasian adults and children. Thorax, 1984. 39: 535-538.
9. ALLEN S.M., HUNT B., GREEN M.: Fall in vital capacity with posture. Brit. J. Dis. Chest 1985. 79: 267-271.

A. Grassino - C. Fracchia - C. Rampulla - L. Zocchi
Respiratory Muscles in COPD
Bi & Gi Publishers, Verona - Springer-Verlag, London, 1988

Intrinsic PEEP and its ramifications in patients with respiratory failure

J. Milic-Emili, S. B. Gottfried, A. Rossi
Meakins-Christie Laboratories, McGill University, Montreal, Quebec, Canada

SUMMARY: Mechanically ventilated COPD patients frequently present dynamic hyperinflation and intrinsic positive end-expiratory pressure (PEEPi), which usually goes unrecorded and, as a result, unrecognized.
The condition nevertheless has important clinical ramifications.
Its presence and extent can and should be simply and non-invasively determined at the bedside, with obvious benefits for the management of critically ill patients.

KEY WORDS: Positive end-expiratory pressure (PEEP); positive end-expiratory pressure, intrinsic (PEEPi); respiratory failure; COPD; hyperinflaction dynamic.

Pulmonary hyperinflation, which is defined as a consistent increase in the end-expiratory lung volume above the predicted functional residual capacity (FRC), is a characteristic abnormality in patients with acute or chronic airways obstruction[1]. While this may be the result of increased lung compliance (e.g. pulmonary emphysema), dynamic factors may also be responsible. For example, when there is a significant increase in airway resistance, the rate of lung emptying is unduly slowed and, by necessity, is interrupted by the next inspiratory effort. When the breathing frequency increases, as with exercise or increased ventilatory demands for other reasons, the expiratory time shortens and hence the end-expiratory lung volume may increase above the relaxed FRC position (elastic equilibrium volume), i.e. hyperinflation ensues. This process is referred to as *dynamic hyperinflation*. Expiratory flow may also be retarded by other mechanisms which contribute to the development of dynamic hyperinflation. These include activity of the inspiratory muscles during expiration (post-inspiration inspiratory activity), as well as activation of laryngeal adductor muscles with expiratory narrowing of the glottic aperture[10, 21, 24, 25]. Although the term dynamic hyperinflation suggests the need for active respiratory efforts, this phenomenon may also be present during passive mechanical ventilation[4, 14, 16, 27, 29]. Indeed, the end-expiratory position during mechanical ventilation will exceed the relaxation volume of the respiratory system whenever the time required for passive expiration to proceed to completion (i.e. to the elastic equilibrium volume) is increased disproportionately to the available expiratory duration imposed by the chosen ventilator settings[14, 29]. In the absence of respiratory muscle activity, the rate of passive lung deflation is determined by the balance between the elastic recoil stored during the preceding lung inflation and the opposing total flow-resistance offered by the respiratory system (including endotracheal tube, ventilator and additional equipment). Accordingly, the stiffer the respiratory system (i.e. decreased compliance), the quicker will be the rate of lung emptying; conversely, increased flow-resistance impedes the rate of lung deflation. Accordingly, as a general rule, the magnitude of dynamic hyperinflation will be proportional to the tidal volume and the mechanical time constant of the respiratory system (i.e. the product of total respiratory resistance and compliance) and inversely related to expiratory duration[18, 35].

In patients with advanced chronic obstructive pulmonary diseases (COPD) mechanically ventilated for management of acute respiratory failure, dynamic hyperinflation is almost invariably present because the time-course of passive expiration is prolonged by increased respiratory resistance[14, 27, 29]. Thus, when airway obstruction worsens or expiratory time decreases with increased ventilatory demands, the dynamically determined FRC will rise. It should be noted, however, that dynamic hyperinflation is not restricted to patients with endstage COPD and expiratory airflow limitation. Respiratory resistance may be increased for a number of reasons in critically ill, intubated and mechanically ventilated patients[17, 34]. For example, the added flow-resistance due to a narrow bore endotracheal tube can be substantial[2, 14] and can be very effective in retarding expiratory flow, thereby contributing to dynamic hyperinflation. Tachypnea with a concomitant reduction in expiratory duration will also have a similar effect[4, 14, 27, 29].

The presence of dynamic hyperinflation implies that alveolar pressure remains positive throughout the expiratory time. In effect, this is equivalent to the deliberate use of PEEP and has been referred to as 'intrinsic' or 'auto' PEEP[15, 17, 29]. In contrast with positive end-expiratory pressure set by the ventilator, however, intrinsic PEEP (PEEPi) is not recorded by the ventilator pressure manometer. As a result, PEEPi is frequently not recognized and has led to use of the term 'occult' PEEP[29]. This is not surprising, in that the ventilator manometer measures the pressure at the airway opening (Pao) relative to atmosphere - and not alveolar pressure. To the extent that end-expiratory lung volume is increased and results in continued expiratory flow, then alveolar pressure must exceed the measured airway pressure. This is in contrast with mechanical PEEP devices which in general provide a constant applied pressure at the airway opening, independent of expiratory flow. This problem is perhaps best illustrated in patients with advanced COPD and dynamic airway compression, where the ventilator manometer is downstream from the site of flow limitation and gives no indication of the existing upstream alveolar pressure proximal to the point of flow limitation[27].

In clinical practice, PEEPi should be suspected whenever one notices that the ventilator spirometer fills continuously throughout expiration only to be abruptly interrupted by the onset of the next ventilator breath. As suggested by Pepe and Marini[27], the presence of PEEPi can be confirmed and its magnitude directly measured by simply occluding the airway opening at end-expiration immediately prior to the onset of the subsequent mechanical inflation. Under these conditions, i.e. absence of expiratory flow with the airway occluded, alveolar pressure will be reliably transmitted to the airway opening and PEEPi will be accurately recorded by the ventilator manometer[27, 29].

As recently demonstrated by Rossi and colleagues[29], PEEPi may also be estimated from continuous recordings of Pao and flow during mechanical ventilation. In absence of PEEPi, expiratory flow ceases within the expiratory time set on the ventilator and inspiratory flow begins symultaneously with the onset of positive pressure applied for mechanical lung inflation. In the presence of dynamic hyperinflation, however, expiratory flow is present throughout expiration and inspiratory airflow cannot begin simultaneously with the onset of the next ventilator breath. The pressure which must be applied by the ventilator before inspiratory flow begins represents the positive pressure required to counterbalance the elastic recoil present at the dynamically increased end-expiratory lung volume, i.e. PEEPi. With mechanical ventilators that provide on-line measurements of airway pressure and flow, PEEPi can be readily and continually determined at the bedside. It should be noted that despite the different approaches, values of PEEPi measured in this way are comparable to results obtained with the end-expiratory occlusion method of Pepe and Marini[27]. Other methods which entail more complex analysis of either expiratory volume-pressure or volume-flow relationships have also been used to determine the presence and magnitude of PEEPi[14, 24, 30].

What are the clinical implications of PEEPi in the mechanically ventilated patient? Static compliance of the total respiratory system (Crs) is frequently measured in the ICU in order to assess respiratory mechanical function and aid in clinical decision-making[5, 33]. This is conventionally determined from single measurements of volume and pressure during airway occlusions performed at end-inflation. Crs is then computed by dividing the tidal volume (VT) by the difference between the 'plateau' in airway pressure during end-inspiratory airway occlusion (P plateau) plus the level of PEEP set by the ventilator (if any), i.e. Crs = VT/(Pplateau + PEEP). This approach assumes that the end-expiratory lung volume during mechanical ventilation corresponds to the relaxed FRC position of the respiratory system, or is at least increased by a predicted amount with applied PEEP. In the presence of dynamic hyperinflation and PEEPi, clearly this assumption is false and will result in systematic underestimation of Crs. Taking PEEPi into account, Rossi et al.[29] have found underestimations of Crs of up to 48%. In fact, such errors were not uncommon and occurred in 10 of 14 patients studied. This problem can be easily overcome by routinely performing airway occlusions at the end of passive expiration in addition to end-inflation. The airway pressure recorded during end-expiratory occlusion will indicate the true magnitude of positive end-expiratory pressure, whether set by the ventilator or PEEPi resulting from dynamic hyperinflation. Crs may then be accurately determined as the tidal volume divided by the difference in airway pressure between end-inspiratory and end-expiratory occlusions. Clearly, commercial ventilators should be provided with the capability of performing not only automatic end-inspiratory airway occlusions, as is at present the case, but also automatic end-expiratory occlusions.

It is generally assumed that patient effort is minimal during 'assisted' mechanical ventilation and as a result the work of breathing is largely taken over by the ventilator[20]. This assumption is true provided that only a modest inspiratory effort is needed or 'triggers' assisted ventilator breaths. Under usual conditions this requires that the ventilated patient reduce airway pressure below the chosen 'sensitivity' level, generally set 1-5 cm H_2O below the prevailing end-expiratory pressure. In the presence of PEEPi, however, the patient must generate a negative inspiratory pressure, equivalent in magnitude to the opposing elastic recoil pressure plus the triggering pressure before an assisted ventilator breath can be initiated. Clearly, the inspiratory efforts performed by the patient during 'assisted' mechanical ventilation cannot be assumed to be negligible in the presence of PEEPi. In fact, we have seen patients who had to generate active inspiratory pressures greater than 20 cm H_2O. This has great importance in the management of ICU patients because the large inspiratory efforts due to PEEPi, associated with decreased inspiratory muscle strength (implicit with hyperinflation), can lead to inspiratory muscle fatigue[3].

Pepe and Marini[27] have recognized and discussed the hemodynamic consequences of PEEPi in mechanically ventilated patients. No different from externally applied PEEP set on the ventilator, the increased intrathoracic pressure present with significant levels of PEEPi will impede venous return, reduce cardiac output, and to some extent be transmitted to the intrathoracic vasculature. This will introduce errors in the measurement and interpretation of central hemodynamic pressure recordings. Failure to recognize the presence of PEEPi in this setting may result in inappropriate fluid restriction and unnecessary vasopressor therapy[27, 29]. Moreover, the adverse hemodynamic effects of PEEPi may in fact be greater than comparable amounts of externally applied PEEP in patients with severe COPD. First, a given amount of end-expiratory pressure will most likely be associated with a higher mean intrathoracic pressure averaged over the course of expiration[9]. Second, unlike patients with acute respiratory distress syndrome in whom PEEP therapy is most frequently utilized, increased compliance in patients with pulmonary emphysema will permit a relatively greater fraction of alveolar pressure to be transmitted to the intrathoracic vasculature[8, 27].

While the presence of dynamic hyperinflation and PEEPi has significant implications for the management of the mechanically ventilated patient, it is also of considerable importance when weaning from ventilatory support is being attempted. Patients with acute respiratory failure almost invariably exhibit rapid, shallow respirations during periods of spontaneous breathing[13, 26]. To the extent that respiratory frequency increases, and hence expiratory time is decreased, this will result in the development of dynamic hyperinflation and increasing PEEPi. In fact, in a study on 14 spontaneously breathing patients with COPD in acute respiratory failure, PEEPi was found in each individual and averaged approximately 9 cm H_2O[13, 26]. The presence of PEEPi in the spontaneously breathing patient implies that, in addition to the pressure needed to produce the actual breathing movements, the inspiratory muscles are required to generate sufficient inspiratory force in order to overcome the opposing positive recoil pressure (i.e., PEEPi) before inspiratory airflow will begin. In this respect, PEEPi represents an *inspiratory threshold load*[7, 23]. This additional pressure requirement places a significant burden on the inspiratory muscles, whose performance as pressure generators is already impaired because of pulmonary hyperinflation and other factors[15, 28]. As a maximum inspiratory pressure of 20-30 cm H_2O is generally taken as indicating sufficient respiratory muscle strength to enable spontaneous ventilation in such patients[11, 31], the additional mechanical load which PEEPi imposes upon the inspiratory muscles is indeed considerable. This is particularly important when considering that the increased end-expiratory lung volume will necessarily reduce the inspiratory muscle strength, increase the mechanical work and oxygen cost of breathing, and decrease the respiratory muscle blood flow and energy supply[6, 12, 13, 19, 28]. This should also predispose to inspiratory muscle fatigue[3, 28]. Clearly, the presence of high levels of PEEPi should herald difficulty in weaning the patient from mechanical ventilation.

As mentioned above, maximum inspiratory pressure is commonly used as a criterion for weaning[11, 31]. In the presence of dynamic hyperinflation, maximum inspiratory pressure will decrease in a predictable fashion with increasing lung volume, related in part to the force-length properties of the inspiratory muscles. Under these conditions, however, maximum inspiratory pressure does not represent the *total* pressure actually developed by the inspiratory muscles, but is underestimated because of the opposing positive elastic recoil pressure present at lung volumes above the relaxed FRC, i.e. PEEPi. Nevertheless, the maximum inspiratory pressure measured in this fashion is clinically relevant to the extent that it does in fact represent the *net* pressure available to produce respiratory movements and airflow. While changes in maximum inspiratory pressure may be due to alterations in lung volume rather than intrinsic respiratory muscle function, it accurately reflects the extent of the reduction in effective ventilatory capacity of the inspiratory muscles. In this regard, with extreme hyperinflation, mechanical ventilation may be achieved at lung volumes exceeding the voluntary total lung capacity[16, 32]. This condition, which perhaps is not uncommon, indicates inability to generate sufficient *net* inspiratory muscle force to provide for spontaneous respiration, and hence complete ventilator dependence must ensue. It is apparent that mechanical ventilation cannot be discontinued until lung volume sufficiently decreases (because of reduced airways obstruction, alveolar dead space, or minute ventilation) or intrinsic inspiratory muscle function itself improves significantly for other reasons.

Fundamental in the approach to the mechanically ventilated patient in whom dynamic hyperinflation and PEEPi occur is the need to recognize if PEEPi is in fact present. Indeed, we believe that measurement of PEEPi should become a part of routine ventilator monitoring in mechanically ventilated patients, particularly those with airways obstruction. This will allow for reliable measurement and interpretation of other frequently determined cardiopulmonary variables, such as respiratory system compliance, pulmonary capillary wedge pressure, and maximum inspiratory pressure. The potential adverse effects of PEEPi require that, in addition, management should be specifically directed towards those factors contributing to the development of PEEPi. This includes medical therapy aimed at reducing the severity of airflow obstruction as well as excessive minute ventilation (due to fever, metabolic acidosis, inadequate pain relief, etc.). The inspiratory flow setting should be adjusted to maximize the time provided for passive expiration to occur. While in theory a reduction in tidal volume should be beneficial, for a given minute ventilation this would require an increased respiratory frequency (and therefore a decreased expiratory time), so that it is unlikely that the level of PEEPi would be appreciably altered[29].

It is important to note that the magnitude of PEEPi can be influenced by externally applied PEEP. In fact, Rossi et al.[29] have demonstrated that expiratory flow limitation due to dynamic airway compression in mechanically ventilated COPD patients results in high levels of PEEPi. Under these conditions, an increase in downstream impedance (i.e. externally applied PEEP) relative to the site of expiratory flow limitation should have little effect on expiratory flow until the applied PEEP exceeds the level of PEEPi. This implies that within these limits, PEEPi should dcrease with increasing externally applied PEEP, i.e. PEEPi is replaced by externally applied PEEP. Use of continuous positive airway pressure (CPAP) may be useful in this setting to facilitate weaning from ventilatory support[22, 29]. In such patients, CPAP should decrease PEEPi without affecting expiratory flow. As a result, less pressure would be wasted by the inspiratory muscles in overcoming PEEPi and more would be available to produce the actual breathing movements. In fact, it is conceivable that the early application of CPAP in some acutely ill patients with chronic airways obstruction may avoid the need for mechanical ventilation entirely[29].

In summary, the presence of dynamic hyperinflation and PEEPi is a common and yet too often unrecognized occurrence in mechanically ventilated patients, particularly those with airways obstruction. The presence of PEEPi has important clinical ramifications. Simple methods are available for the rapid, non-invasive determination of the magnitude of PEEPi at the bedside, using presently available equipment. Prompt recognition of PEEPi will allow for proper measurement and interpretation of a variety of cardiopulmonary variables, and improved clinical decision-making and management of critically ill patients.

References

1. BATES D.V., MACKLEM P.T., CHRISTIE R.V. ed.. Respiratory function in disease. Philadelphia P.A., W.B. Saunders Company, 1971.

2. BEHRAKIS P.K., HIGGS B.D., BAYDUR A., ZIN W.A., MILIC-EMILI J.. Respiratory mechanics during halothane anesthesia and anesthesia-paralysis in man. J. Appl. Physiol. 1983. 55: 1085-92.

3. BELLEMARE F., GRASSINO A.. Force reserve of the diaphragm in patients with chronic obstructive pulmonary diesease. J. Appl. 1983. 55: 8-15.

4. BERGMAN N.A.. Intrapulmonary gas trapping during mechanical ventilation at rapid frequencies. Anesthesiology 1972. 37: 626-33.

5. BONE R.C.. Diagnosis of causes for acute respiratory distress by pressure-volume curves. Chest 1976. 70: 740-6.

6. BUCHLER B., MAGDER S., KATSARDIS H., JAMMES Y., ROUSSOS C.. Effects of pleural pressure and abdominal pressure on diaphragmatic blood flow. J. Appl. Physiol. 1985. 58: 691-97.

7. CAMPBELLE E.J.M., DICKINSON C.J., DINNICK O.P., HOWELL J.B.L.. The immediate effects of threshold loads on the breathing of men and dogs. Clin. Sci. 1961. 21: 309-20.

8. CHAPIN J.C., DOWNS J.B., DOUGLAS M.E., MURPHY E.J., RUIZ B.C.. Lung expansion, airway pressure transmission, and positive end-expiratory pressure. Arch., Surg, 1979. 114: 1193-7.

9. COLGAN F.J., BARROW R.E., FANNING G.L.. Constant positive-pressure breathing and cardiorespiratory function. Anesthesiology 1971. 34: 145-51.

10. COLLETT P.W., BRANCATISANO T., ENGEL L.A. Changes in the glottic aperture during bronchial asthma. Am. Rev. Respir. Dis. 1983. 128: 719-23.

11. FEELEY T.W., HEDLEY-WHITE J.. Weaning from controlled ventilation and supplemental oxygen. New. Engl. J. Med. 1975. 292: 903-906.

12. FIELD S., KELLY S.M., MACKLEM P.T.. The oxygen cost of breathing in patients with cardiorespiratory failure. Am. Rev. Respir. Dis. 1982. 126: 9-13.

13. FLEURY B., MARCIANO D., TALAMO C., AUBIER M., PARIENTE R., MILIC-EMILI J.. Work of breathing in patients with chronic obstructive pulmonary disease in acute respiratory failure. Am. Rev. Respir. Dis. 1985. 131: 822-27.

14. GOTTFRIED S.B., ROSSI A., HIGGS B.D., CALVERLEY P.M.A., ZOCCHI L., BOZIC C., MILIC-EMILI J.. Non invasive determination of respiratory system mechanics during mechanical ventilation for acute respiratory failure. Am. Rev. Respir. Dis. 1985. 131: 414-20.

15. KELLY S.M., ROSA A., FIELD S., COUGHLIN M., SHIZGAL H.M., MACKLEM P.T.. Inspiratory muscle strength and body composition in patients receiving total parenteral nutrition therapy. Am. Rev. Respir. Dis. 1984. 130: 33-37.

16. KIMBALL W.R., LEITH D.E., ROBINS A.G. Dynamic hyperinflation and ventilator dependence in chronic obstructive pulmonary disease. Am. Rev. Respir. Dis. 1982. 126: 991-5.

17. LAVIETES M.H., ROCHESTER D.F.. Assessment of airway function during assisted ventilation. Lung 1981. 159: 219-29.

18. LESOUEF P.N., England S.J., BRYAN A.C.. Passive respiratory mechanics in new-borns and children. Am. Rev. Respir. DIs. 1984. 129: 552:56.

19.MACKLEM P.T. Hyperinflation. Am. Rev. Respir. Dis. 1984. 129: 1-2.

20. MARINI J.J., CAPPS J.S., CULVER B.H.. The inspiratory work of breathing during assisted mechanical ventilation. Chest 1985. 87: 612-18.

21. MARTIN J., POWELL E., SHORE S., EMRICH J., ENGEL L.A.. The role of the respiratory muscles in the hyperinflation of bronchial asthma. Am. Rev. Respir. Dis. 1980. 123: 441-7.

22. MARTIN J.G., SHORE S., ENGEL L.A.. Effect of continuous positive airway pressure on respiratory mechanics and pattern of breathing in induced asthma. Am. Rev. Respir. Dis. 1982. 126: 812-17.

23. MEAD J.. Responses to loaded breathing. Bull. Eur. Physiopath. Respir. 1979. 15: 61-71 (Suppl.).

24. MORTOLA J.P., MILIC-EMILI J., NOVORAJ A., SMITH B., FOX G., WEEKS S.. Muscle pressure and flow during expiration in infants. Am. Rev. Respir. Dis. 1984. 129: 49-53.
25. MULLER N., BRYAN A.C., ZAMEL N.. Tonic inspiratory muscle activity as a cause of hyperinflation in histamine-induced asthma. J. Appl. Physiol. 1980. 49: 869-74.
26. MURCIANO D., AUBIER M., BUSSI S., DERENNE J-PH., PARIENTE R., MILIC-EMILI J.. Comparison of esophgeal, tracheal, and mouth occlusion pressure in patients with chronic obstructive pulmonary disease during acute respiratory failure. Am. Rev. Respir. Dis. 1982. 126: 837-41.
27. PEPE P.E., MARINI J.J.. Occult positive end-expiratory pressure in mechanically ventilated patients with airflow obstruction. Am. Rev. Respir. Dis. 1982. 126: 166-70.
28. ROCHESTER D.J., ARORA N.S.. Respiratory muscle failure. Med. Clin. N. Amer. 1983. 67: 573-97.
29. ROSSI A., GOTTFRIED S.B., ZOCCHI L., HIGGHS B.D., LENNOX S., CALVERLEY P.M.A., BEGIN P., GRASSINO A., MILIC-EMILI J.. Measurement of static compliance of the total respiratory system in patients with acute respiratory failure during mechanical ventilation: the effect of intrinsic positive end-expiratory pressure. Am. Rev. Respir. Dis. 1985. 131: 672-7.
30. SAETTA M., ROSSI A., GOTTFRIED S.B., LAPORTA D., ZOCCHI L., BEGIN P., MILIC-EMILI J.. Expiratory volume-flow relationship during mechanical ventilation in patients with acute respiratory failure. Am. Rev. Respir. Dis. 1985. 131: A132.
31. SAHN S.A., LAKSHMINARAYAN S.. Bedside criteria for discontinuation of mechanical ventilation. Chest 1973. 63: 1002-5.
32. SHARP J.T., VAN LITH P., NUCHPRAYOON C.V., BRINEY R., JOHNSON F.N.. The thorax in chronic obstructive lung disease. Am. J. Med. 1968. 44: 39-46.
33. SUTER P.M., FAIRLEY H.B., ISENBERG M.D.. Optimum end-expiratory airway pressure in patients with acute pulmonary failure. N. Engl. J. Med. 1975. 292: 284-9.
34. SYBRECHT G.W., TAUBNER E.M., BOHM M.M., FABEL H.. Mechanical properties of the respiratory system and mouth-occlusion pressure in patients acutely intoxicated with hypnotics. Lung. 1979. 156: 49-56.
35. WINEGAR E., SINNETT E.E., LEITH D.E.. Dynamic mechanisms determine functional residual capacity in mice, *Mus musculus*. J. Appl. Physiol. 1979. 46: 867-71.

A. GRASSINO - C. FRACCHIA - C. RAMPULLA - L. ZOCCHI
Respiratory Muscles in COPD
Bi & Gi Publishers, Verona - Springer-Verlag, London, 1988

Early Changes in respiratory mechanics in acute respiratory failure

A. ROSSI[2], R. POGGI[2], E. MANZIN[1], C. BROSEGHINI[2], R. BRANDOLESE[1]
[1] *Department of Anesthesia and Intensive Care, City Hospital, Padua, Italy*
[2] *Institute of Occupational Health, University of Padua, Padua, Italy*

SUMMARY: The purpose of this study was to investigate early changes in respiratory mechanics in mechanically ventilated patients with acute respiratory failure (ARF), using modern ventilators equipped with operational and measuring devices. To this end, 24 patients with ARF related to acute exacerbation of chronic airway obstruction (CAO, 8 patients) or pulmonary edema, both cardiogenic (8 patients) and noncardiogenic (8 patients with ARDS adult respiratory distress syndrome), were examined within 24 hours of the onset of mechanical ventilation. We found that: 1. patients with CAO were characterized by high 'intrinsic' PEEP (PEEPi), mean 13.6(6.7) cm H_2O, relatively low compliance, increased respiratory resistance, both maximum and minimum, with marked frequency-dependence of resistance; 2. in pulmonary edema PEEPi was present, amounting, on average, to 3.8 and 3.0 cm H_2O in non-cardiogenic and cardiogenic pulmonary edema respectively, and respiratory resistance was increased, though exhibiting a significant frequency-dependence only in the non-cardiogenic type of edema; 3. early ARDS was characterized by low compliance, mean 0.035(0.005) L/cm H_2O, and frequency-dependence of resistance. We conclude that changes in respiratory mechanics in the mechanically ventilated patient with ARF should be assessed to gain a better understanding of the patient's condition, and that noninvasive study can be easily performed with modern operational and measuring ventilators.

KEY WORDS: Respiratory mechanics; respiratory failure, acute; chronic airway obstruction CAO; PEEP; PEEPi; cardiogenic pulmonary edema; chronic obstructive pulmonary disease (COPD).

Introduction

It is a commonly accepted notion that assessment even of appreciable changes in respiratory mechanics in patients with acute respiratory failure (ARF) during mechanical ventilation might not be a simple task in the 'crowded' atmosphere characterized by end-inspiratory static compliance. In fact, 'whereas the routine use of intra-arterial monitoring lines and flow directed pulmonary artery catheters has made repeated measurements of arterial and mixed venous blood gas tensions, intrapulmonary shunt, right and left ventricular filling pressures and cardiac output common in the intensive care setting, results of other pulmonary function tests are quite difficult to obtain under usual clinical circumstances'[1]. In other words, sophisticated and invasive measurement of cardiovascular function by means of a

Swan-Ganz catheter is more familiar in the intensive care unit (ICU) than simple and noninvasive determination of changes in respiratory mechanics. It is therefore not surprising that few data on respiratory mechanics in patients with early ARF during mechanical ventilation have been reported in literature, which is, on the other hand, saturated with physiological measurements in chronic stable patients.

By contrast, the airways of a mechanically ventilated patient are ideal for measuring changes in the passive mechanical properties of the respiratory system with simple, commonly available equipment, i.e. a pneumotachograph for measurement of respiratory flow and change in lung volume, and a differential pressure transducer for measurement of pressure at the airway opening (the proximal end of the endotracheal tube)[2]. This is due to the fact that the upper airway is bypassed by the endotracheal tube and that respiratory muscles are often relaxed or not active during mechanical ventilation[3]. Besides, it has to be mentioned not only that simple manouvers, namely end-inspiratory and end-expiratory brief (about 1 s) occlusions at the airway opening, are required to obtain a detailed analysis of the mechanical properties of the respiratory system during mechanical ventilation[4,5], but also that some modern mechanical ventilators, e.g. the 900C Servo[6], are already equipped for such manouvers and on-line non-invasive measurements.

This study was undertaken to investigate, non invasively, early changes in the mechanical properties of the total respiratory system in patients with ARF related to pulmonary edema or acute airway obstruction, within 24 hours of the onset of mechanical ventilation, using the operational and measuring devices provided by the ventilator.

Methods

Twenty-four patients admitted to the intensive care unit (ICU) at the City Hospital, Padua, with ARF related to pulmonary edema or acute airway obstruction were examined. Characteristics of the patients are reported in Table 1. The diagnosis of pulmonary edema was made on the basis of clinical signs and the chest X-ray on admission to the ICU. Acute airway obstruction was due to a severe exacerbation of a pre-existing and well documented chronic airway obstruction (CAO), whith bronchial asthma in two patients and chronic obstructive pulmonary disease (COPD) in the other six. It can also be seen in Table 1 that the cause of pulmonary edema was cardiogenic in 8 patients (CPE: cardiogenic pulmonary edema) and non-cardiogenic in the other 8 patients. The latter fullfilled the criteria for the diagnosis of the Adult Respiratory Distress Syndrome (ARDS)[1,7]. All patients were intubated (Portex cuffed endotracheal tube, ETT, internal diameter ranging from 7 to 8.5 mm), and mechanically ventilated with intermittent mandatory ventilation (IMV), using a Servo 900 C Siemens ventilator, with constant inspiratory flow ($\dot{V}I$). The mode and the patterns of mechanical ventilation were established on the basis of the clinical judgement of the physicians in charge. Informed consent to the study was obtained by the next of kin of the patients. The research protocol was approved by the ethical

Table 1. Patient characteristics

PATIENTS	SEX	AGE (yr)	CLINICAL DIAGOSIS
Group A: CPE			
1	F	80	Inferior myocardial infarction
2	M	61	Myocardial infarction
3	M	43	Anterior myocardial infarction
4	F	59	Mitral stenosis
5	F	65	Anterior myocardial infarction
6	M	58	Systemic hypertension
7	M	65	Myocardial infarction
8	F	50	Renal hypertension
Group B: ARDS			
9	F	63	Ab ingestis pneumonia
10	F	65	Hypovolemic shock
11	M	23	Scalded skin syndrome
12	M	18	Shock lung
13	F	46	Ab ingestis pneumonia
14	M	38	Polytrauma
15	M	70	Peritonitis, septicemia
16	M	40	Pneumonitis, sepsis
Group C: CAO			
17	F	17	Bronchial asthma
18	F	84	COPD
19	F	66	COPD
20	M	79	COPD
21	M	65	COPD
22	M	77	COPD
23	F	36	Bronchial asthma
24	M	73	COPD

Abbreviations: CPE = cardiogenic pulmunary edema; ARDS = adult respiratory distress syndrome; CAO = cronic airway obstruction; COPD = chronic obstructive pulmonary disease.

authorities of the Hospital. Patients were sedated (morphine) and some of them (all the patients with ARDS, with one exception, no. 12) were paralyzed (pancuronium bromide), again at the discretion of the departmental physicians. No patient was sedated, paralyzed or subjected to ventilatory pattern changes because of our experimental protocol. All patients were examined before any positive end-expiratory pressure (PEEP) was set by the ventilator, i.e. on ZEEP (zero end-expiratory pressure).

Airway pressure (Pao) and respiratory flow were measured by the pressure transducers included in the Servo 900C; expired lung volume was obtained by electrical integration of the flow signal[6]. All the signals were recorded on a multichannel pressure ink pen recorder (Mingograph Siemens), at a paper speed of 15.5 or 31 mm/s. A representative record is shown in Figure 1.

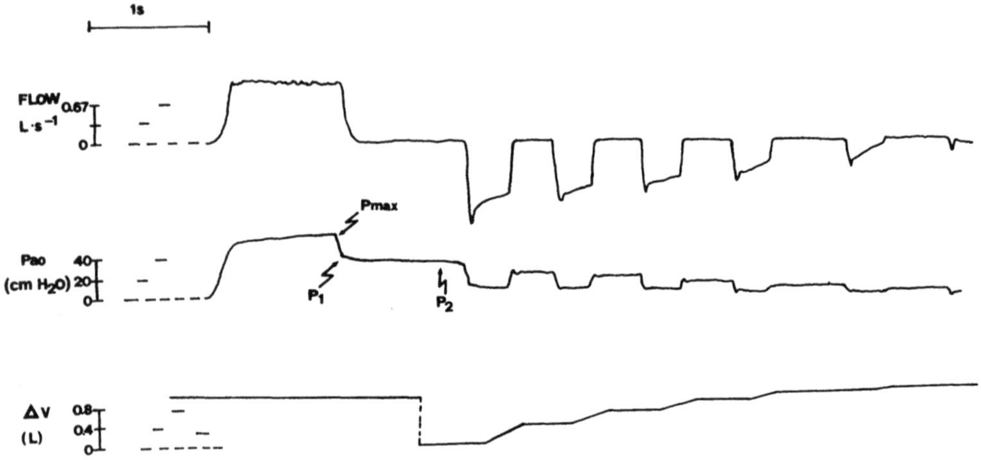

Fig. 1. - Records of respiratory flow, airway pressure (Pao), and expired volume (ΔV) in a mechanically ventilated patient with ARDS (no.11). When airway occlusion is performed at the end of the mechanical inflation, Pao drops from its maximal value, Pmax, to P1, and then slowly decays to a plateau (P2). After the plateau was observed, the occlusion was released and brief interruptions were performed throughout the relaxed expiration[11].

Experimental Procedure and Data Analysis

Regular ventilator setting was recorded without any change for several breaths, then the airway opening was occluded at the end of a mechanical inflation by means of the end-inspiratory hold of the ventilator. The occlusion was kept until, after the initial drop (P1), a plateau in Pao (P2) indicated equilibration between alveolar and mouth pressure (i.e. the proximal end of the ETT). As soon as the plateau in Pao was observed, the occlusion was released for relaxed expiration[8]. Under these circumstances the plateau in Pao represented the elastic recoil of the total respiratory system at the end of mechanical inflation[8]. After ten regular breaths with end-inspiratory occlusion, the airway was also occluded at the end of a tidal expiration, using the end-expiratory hold of the ventilator, for direct measurement of 'intrinsic' PEEP, when present[9].

The static and dynamic compliances of the total respiratory system (Cst,rs and Cdyn,rs) were computed as the ratio of the expired tidal volume (VT) and the difference between P2 and P1, respectively, to the end-expiratory plateau pressure[2]. In this way PEEPi, when present, was taken into account and the correct value of both compliances was obtained[4]. Total respiratory resistance (Rtot) was computed as the ratio of the difference between maximum Pao (Pmax) and P2 and $\dot{V}I$. According to the analysis by Bates et al.[10], this represented the maximum resistance of the respiratory system. The minimum resistance (Rtot min) was computed from the ratio between (Pmax-P1) and $\dot{V}I$.

The ideal application of the end-inspiration occlusion method requires an instantaneous occlusion of the airway opening, and hence an instantaneous drop of flow to zero. This was never the case in our study (Fig. 1), not only because of the finite occlusion time of the occlusion valve, but also because the end-inspiratory hold of the ventilator, by definition, operates only after mechanical inflation has taken place. This necessarily results in an underestimation of both resistive pressures, namely Pmax-P1 and Pmax-P2. This is because, during the period from the end of inflation, indicated by the rapid decay in Pao and $\dot{V}I$, to the full closure of the valve, flow is still present, and hence the volume of gas slightly increases to a finite amount which is determined by the elastic properties of the respiratory system, namely Cst,rs[5]. The true values of resistance were therefore obtained correcting the measured values according to the equatien:

Rtot (Rtot min) = Rtot' (Rtot min') + (ΔV/Cst,rs) / $\dot{V}I$[1],

where Rtot' (Rtot min') are the values of resistance recorded before correction and ΔDV is the inflated volume during the occlusion of the valve[11].

Maximum and minimum intrinsic respiratory resistance (Rrs and Rrs min, respectively), were obtained by subtracting from the total the resistance of the endotracheal tubes and the inspiratory line of the ventilator. The resistances of the ETT were computed according to the equation $P = a\,\dot{V}^b$, where a is the value of P at $\dot{V} = 1$ L/s and b is a dimensionless index describing the shape of the curve[5, 12]. The resistance of the inspiratory line of the ventilator was measured directly in each study, by disconnecting the patient from the inspiratory line for a few inflations. Mean values of three measurements were used for each variable.

Statistical analysis was performed using the one way analysis of variance (ANOVA), and regression analysis with the least square method; a $p < 0.05$ was accepted as significant.

Results and discussion

Individual and average (SD) results of measurements of respiratory mechanics are reported in Table 2. PEEPi was present, with few exceptions, in almost all the patients. As expected on the basis of previous reports[4, 6, 9], PEEPi was significantly higher in patients with acute exacerbation of CAO ($F = 14.5$; $p < 0.01$), averaging 13.6(6.7) cm H_2O, but it must be emphasized that the end-expiratory pressure was also slightly positive in most patients with pulmonary edema, both cardiogenic and noncardiogenic, averaging 3.0 (2.1) and 3.8 (3.0) cm H_2O respectively, even if mechanical ventilation was taking place from ZEEP. To our knowledge, apart from sporadic observations[4], this is the first systematic report on 'intrinsic' PEEP, i.e. PEEPi, in mechanically ventilated patients with pulmonary edema. The ventilatory frequency, I:E ratio, and VT were not significantly different between the three groups of patients and were in accordance with widely used ventilator settings.

The hemodynamic effect of PEEP could be appreciable, because of its magnitude, in patients with CAO, as already pointed out by Pepe and Marini[9], while it is probably negligible in mechanically ventilated patients with pulmonary edema both because of the relatively low values and because an external PEEP is frequently set

Table 2. Respiratory mechanics

PATIENTS	PEEPi (cmH20)	Cdyn, rs (L/cmH20)	Cst, rs	Rtot (cmH20/L.s)	Rtot (min)	Rrs	Rrs (min)
Group A: CPE							
1	0.5	0.052	0.055	13.7	12.6	8.0	6.9
2	2.0	0.028	0.038	25.8	21.9	9.4	5.4
3	1.1	0.046	0.056	29.1	24.9	14.3	10.2
4	3.0	0.037	0.044	33.0	28.7	15.7	12.3
5	1.4	0.026	0.032	32.0	25.8	15.4	9.3
6	6.0	0.044	0.046	16.6	15.4	5.4	4.2
7	4.0	0.026	0.040	35.5	28.9	21.5	15.0
8	6.0	0.038	0.043	21.5	17.9	6.8	3.2
mean	3.0	0.037	0.044	25.9	22.0	12.1	8.3
(SD)	(2.2)	(0.010)	(0.008)	(8.0)	(6.1)	(5.5)	(4.1)
Group B: ARDS							
9	5.6	0.025	0.032	30.1	21.9	16.3	8.1
10	2.0	0.023	0.037	33.3	23.3	17.9	7.9
11	8.1	0.029	0.034	28.6	22.9	12.6	6.9
12	3.6	0.028	0.034	37.8	32.0	23.5	17.7
13	4.9	0.036	0.046	27.7	21.7	11.3	5.3
14	5.8	0.025	0.030	29.9	20.9	18.4	10.9
15	0.0	0.026	0.030	20.9	14.9	9.2	3.2
16	0.0	0.028	0.036	26.5	16.6	14.6	4.2
mean	3.8	0.028	0.035	29.4	21.,8	15.5	8.0
(SD)	(2.9)	(0.004)	(0.005)	(4.9)	(5.1)	(4.6)	(4.6)
Group C: CAO							
17	2.8	0.040	0.047	33.4	22.4	21.6	10.6
18	16.0	0.028	0.042	41.7	31.7	27.5	17.5
19	14.0	0.034	0.051	31.7	20.1	17.5	6.0
20	11.0	0.035	0.046	39.7	30.3	26.1	15.9
21	14.0	0.044	0.060	41.8	34.3	28.4	20.9
22	7.0	0.050	0.068	29.7	23.4	16.5	10.2
23	22.0	0.022	0.035	70.6	46.8	57.3	33.5
24	22.0	0.069	0.096	29.9	23.9	16.5	10.5
mean	13.6	0.040	0.056	39.8	29.1	26.4	15.6
(SD)	(6.7)	(0.015)	(0.019)	(13.4)	(8.7)	(13.4)	(8.7)

Abbreviations: PEEPi = "intrinsic" PEEP; Cdyn, rs = dynamic respiratory compliance; Cst, rs = static respiratory compliance; Rtot and Rtot (min) = maximum and minimum respiratory resistance; Rrs and Rrs (min) = maximum and minimum intrinsic respiratory resistance; Rrs and Rrs (min) = maximum and minimum intrinsic respiratory resistance, i.e. after substraction of the resistance of the endotracheal tubes and the inspiratory line of the ventilator. All values of compliances and resistances are in L/cmH_2O and cmH_2O/L.s, respectively.

by the ventilator for therapeutical purposes in such patients. However, it must be stressed that, under these conditions, the computation of Cst,rs often provided by ventilators fitted with an on-line computer may be significantly underestimated, because it does not take PEEPi into account, and that the error may vary with different ventilator settings or/and the patient's respiratory mechanics.

The implications of PEEPi in terms of weaning for CAO patients have been discussed elsewhere[13, 14].

Both Cdyn,rs and Cst,rs were lower than normal in all the patients, with one exception (COPD patient no. 24, Tab. 2)[4], but only Cst,rs was significantly different between the three groups ($F = 5.6$; $p < 0.05$). In patients with acute exacerbation of CAO, a low respiratory compliance was accepted as mainly due to lung hyperinflation[12, 13]. This interpretation is also supported by the high PEEPi, which indicates a large discrepancy between the end-expiratory volume during regular mechanical ventilation and the elastic equilibrium volume of the total respiratory system[4, 12]. In pulmonary edema, the low compliance is determined by accumulation of liquid fluid in the air spaces and changes in the surface forces, rather than by changes in the elastic properties of the chest wall and lung tissue[15, 16, 17]. The lowest mean Cst,rs, 0.035(0.005) L/cm H_2O, probably reflected a more severe air space flooding in patients with ARDS than in cardiogenic pulmonary edema. In this connection it must be stressed that these patients were examined within the first 24 hours of ARF; we therefore suggest that the general notion that compliance is low, but not too low, in the early phase of ARDS could be reconsidered in mechanically ventilated patients[1].

According to the model analysis of the behaviour of the respiratory system with constant flow inflation suggested by Bates et al.[10], the sudden drop from Pmax to P1 (Fig. 1) provides the pressure driving the flow, while the subsequent slow fall from P1 to P2 reflects, during respiratory muscle relaxation, intrapulmonary equilibration between lung units with different time constants, i.e. pendelluft. In this context, Rrs represents the maximum resistance of the total respiratory system, i.e. resistance at zero frequency, while Rrs min is the minimum resistance, i.e. the resistance at infinite frequency: the frequency-dependence of resistance may be represented by the difference between Rrs and Rrs min. In other words, Rrs min is the resistance of the respiratory system in the absence of time constant inequalities within the respiratory system, i.e. the resistance of the conducting network, while Rrs represents the minimum resistance plus the contribution of time constant discrepancies and stress relaxation. In a normal respiratory system, Rrs only slightly exceeds Rrs min for the amount of tissue stress relaxation. By contrast, in respiratory diseases, Rrs can be much larger than Rrs min because of widespread time constant inequalities[2, 5, 18]. In this context an increase in Rrs min reflects increased resistance of the central airway, while a further increase of Rrs reflects the relative contribution of the periphery of the lung.

The frequency-dependence of resistance is a well known phenomenon in patients with stable COPD[18], and was recently demonstrated in mechanically ventilated patients with ARF because of exacerbation of their COPD[5]. Table 2 shows that both Rrs and Rrs min were high in the patients of this group, averaging 26.4(13.4) and 15.6(8.7) cm H_2O/L.s respectively, the Rrs - Rrs min difference amounting to

10.8(5.7) cm H_2O/L.s. However, it also shows that respiratory resistance was high in patients with pulmonary edema, Rrs and Rrs min averaging respectively 12.1(5.5) and 8.3(4.1) cm H_2O/L.s in patients with CPE and 15.1(4.6) and 8.04(4.6) cm H_2/L.s in patients with ARDS. The mean Rrs - Rrs min difference was 3.8(2.0) cm H_2O/L.s in the group of patients with CPE and 7.5(1.9) cm H_2O/L.s in ARDS. Both Rrs and Rrs min were significantly higher in patients with acute exacerbation of their CAO than in the patients with pulmonary edema ($F = 5.9$ and $F = 4.0$ respectively, $p < 0.05$), and the mean Rrs - Rrs min difference, i.e. frequency-dependence of resistance, was also significantly different between the theree groups ($F = 7.5$; $p < 0.01$). In agreement with a previous observation[5], frequency-dependence of resistance was significantly higher ($F = 9.6$, $p < 0.01$) in the patients with acute airway obstruction than in patients with pulmonary edema taken as a whole. However, within this group, frequency-dependence of resistance was more marked in the group with ARDS ($F = 14.6$, $p < 0.01$) than in patients with CPE. These results indicate that the resistance of the central airway may be increased in acute pulmonary edema, but that the periphery of the lung is significantly more involved in permeability than in hydrostatic edema[7]. In this context it has to be repeated that chest X-rays of our patients showed appreciable air space flooding.

The mechanism of increased resistance in pulmonary edema is still controversial[18, 20]. However, a recent quantitative histologic study showed that compression of the airway by the interstitial fluid must be excluded and that other factors should be investigated, such as the vagally-mediated reflexes and presence of airspace luminal edema[21]. As previously said, this study shows that the resistance of the central airway, namely Rrs min, was increased to the same extent in both types of edema, while the maximum respiratory resistance, namely Rrs, was significantly higher in patients with ARDS, resulting in a more marked frequency-dependence of resistance in the latter. This indicated the presence of substantial time constant inequalities within the lung, which were probably caused by discrepancies in elastic properties, as suggested by the low compliance. In fact, as illustrated in Figure 2, a significant correlation was found between Cst,rs and the Rrs - Rrs min difference ($r = 0.64$, $p < 0.01$). This was not the case when the 24 patients were considered together, as shown by Figure 3, where it can be seen that in patients with CAO the Rrs - Rrs min difference was coupled with a relatively high Cst,rs. Clearly, different mechanisms underlie frequency-dependence of resistance in patients with CAO and with ARDS. In this connection it must be reported that in a patient with ARDS, examined in a different situation, several days after the onset of mechanical ventilation Rrs min was low, being 1.8 cm H_2O/L.s, while Rrs was almost three times larger, i.e. 4.8 cm H_2O/L.s[5]. This indicates that, even if resistance of the central airway is normal, there may be frequency-dependence of resistance, which suggests that ARDS may be characterized mainly as peripheral lung disease.

In this context it is appropriate to note the important contribution to total

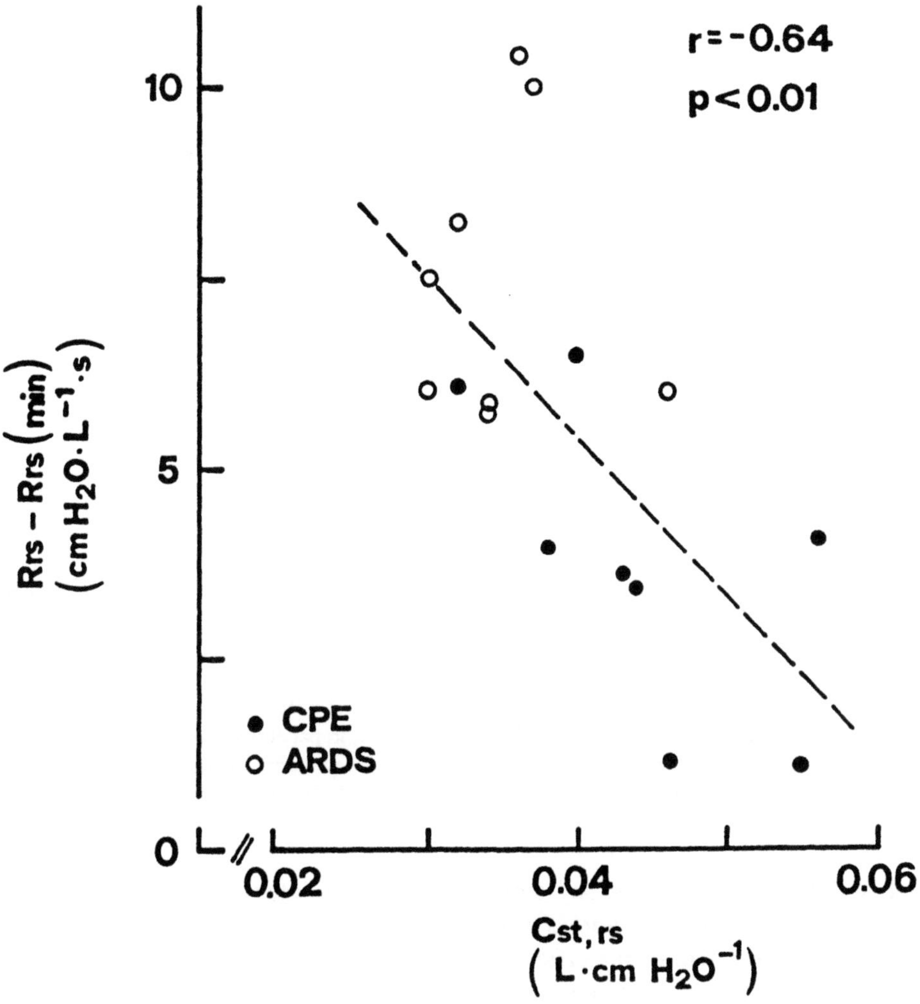

Fig. 2. - Relationship between static respiratory compliance (Cst,rs) and the difference between maximum and minimum intrinsic respiratory resistance, Rrs and Rrs min respectively, in patients with early cardiogenic pulmonary edema (CPE), filled circles, and in patients with adult respiratory distress syndrome (ARDS), empty circles. The dashed line is the regression line.

resistance (Tab. 2) given by the endotracheal tubes and the inspiratory line of the ventilator; this must be kept in mind when evaluating the resistance automatically computed by computer-equipped ventilators for comparison between patients.

Conclusion

Noninvasive but detailed assessment of changes in respiratory mechanics can be easily and readily performed in mechanically ventilated patients with ARF, using the opera-

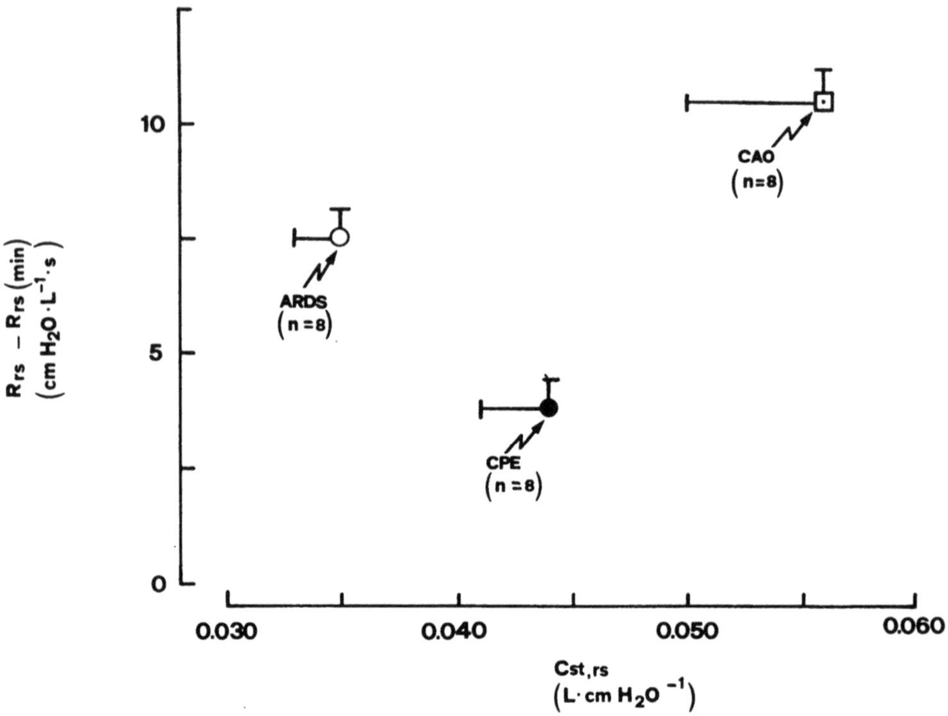

Fig. 3. - Average values of static respiratory compliance (Cst,rs) and difference between maximum and minimum intrinsic respiratory resistance, Rrs - Rrs min, in patients with acute exhacerbation of chronic airway obstruction (CAO), empty square, adult respiratory distress syndrome (ARDS), empty circle, and cardiogenic pulmonary edema (CPE), filled circle. Bars are standard error of the mean.

tional and measuring devices of some modern ventilators from the onset of mechanical ventilation. This allows a better understanding of the disease underlying ARF and assessment of the efficiency of therapeutical manouvers, e.g. PEEP, as well as of the day-by-day evolution of ARF.

We also feel it is important to point out that:

1. 'intrinsic' PEEP is a constant feature in patients with ARF because of exacerbation of CAO, reaching values higher than 20 cm H_2O, but it may also be present in patients with acute pulmonary edema.

2. Respiratory resistance may be high in patients with early pulmonary edema, probably because of airspace flooding and vagally-mediated reflexes, but frequency-dependence of resistance was more marked in patients with ARDS than in patients with CPE, probably because of a more pronounced involvement of the periphery of the lung.

3. Early ARDS, in mechanically ventilated patients, was characterized by a very low respiratory compliance, increased respiratory resistance, and frequency-dependence of resistance.

4. The values of compliance and resistance provided by modern computer-equipped mechanical ventilators must be critically evaluated, since respiratory compliances might be underestimated because of PEEPi, and total resistance includes the significant contribution of the endotracheal tubes and the inspiratory line of the ventilator[22].

Acknowledgements

We are indebted to Prof. J. Milic-Emili for his critical suggestions in the computation of the corrected values of resistances. This work was supported by the National Research Council, Cardiorespiratory Group and Bilateral Project, grants no. 85.00532.04 and 86.00132.04, and the Ministry of Education, Italy. Dr. C. Broseghini holds a fellowship provided by Boehringer Ingelheim spa, Florence, Italy. We wish to thank Mr. F. Bortolami for his helpful technical assistance, and Dr. G. Polese for assistance in data analysis. We also wish to acknowledge the kind and skilful cooperation of the nurses and physicians of the Intensive Care Unit in Padua.

References

1. SUTER P.M.. Assessment of respiratory mechanics in ARDS. In Zapol Z.M., Falke K.J. (Eds.) 'Acute respiratory failure' volume 24, Lung biology in health and disease; New York, Marcel Dekker Inc., 1985: 507-19.
2. MILIC-EMILI J., GOTTFRIED S.B., ROSSI A.. Non-invasive measurement of respiratory mechanics in ICU patients. Int. J. Clin. Monitor Comp., 1986.
3. MILIC-EMILI J.. Measurement of pressure in respiratory physiology. In: Otis AB (ed), Techniques in respiratory physiology - part II, Physiology - vol P4/II, Elsevier, County Clare, Ireland 1984: P412/1-22.
4. ROSSI A., GOTTFRIED S.B., ZOCCHI L., HIGGS B.D., LENNOX S., CALVERLEY P.M.A., BEGIN P. GRASSINO A., MILIC-EMILI J. Measurement of static compliance of total respiratory system in patients with acute respiratory failure during mechanical ventilation. Am. Rev. Respir. Dis. 1985. 131: 672-8.
5. ROSSI A., GOTTFRIED S.B., HIGGS B.D., ZOCCHI L., GRASSINO A., MILIC-EMILI J.. Respiratory mechanics in mechanically ventilated patients with respiratory failure. J. Appl. Physiol., 1985. 58: 1849-58.
6. JONSON B., NORDSTROM L., OLSSON S.G., AKERBACK D.. Monitoring of ventilation and lung mechanics during automatic ventilation. A new device. Bull Physiopath Resp., 1975. 11: 729-43.
7. STAUB N.C.. Pulmonary Edema. Physiol. Rev. 1974. 54: 678-811.
8. ZIN W.A., PENGELLY L.D., MILIC-EMILI J.. Single breath method for measurement of respiratory mechanics in anesthetized animals. J. Appl. Physiol. 1982. 52: 1266-71.
9. PEPE P.E., MARINI J.J.. Occult positive end-expiratory pressure in mechanically ventilated patients with airflow obstruction. Am. Rev. Respir. Dis. 1982. 126:166-70.
10. BATES J.H.T., ROSSI A., MILIC-EMILI J.. Analysis of the behaviour of the respiratory system with constant inspiratory flow. J. Appl. Physiol. 1985. 58: 1840-8.
11. KOCHI T., OKUBO S., ZIN W.A., MILIC-EMILI J.. Flow and volume dependence of pulmonary mechanics in anesthetized cats. J. Appl. Physiol. submitted for publication.

12. GOTTFRIED S.B., ROSSI A., HIGGS B.D., CALVERLEY P.M.A., ZOCCHI L., BOZIC C., MILIC-EMILI J.. Noninvasive determination of respiratory system mechanics during mechanical ventilation for acute respiratory failure. Am. Rev. Respir. Dis. 1985. 131: 414-20.
13. GOTTFRIED S.B., ROSSI A., MILIC-EMILI J.. Dynamic hyperinflation, intrinsic PEEP and the mechanically ventilated patient. Int. and Crit. Care. Med. 1986 in press.
14. FLEURY B., MURCIANO D., TALAMO C., AUBIER M., PARIENTE R., MILIC-EMILI J.. Work of breathing in patients with chronic obstructive pulmonary disease in acute respiratory failure. Am. Rev. Respir. Dis. 1985. 132: 822-7.
15. COOK C.D., MEAD J., SCHREINER G.L., FRANK N.R., CRAIG J.M.. Pulmonary mechanics during induced pulmonary edema in anesthetized dogs. J. Appl. Physiol. 1959. 14: 177-86.
16. GROSSMAN R.F., JONES J.G., MURRAY J.F.. Effects of oleic acid-induced pulmonary edema on lung mechanics. J. Appl. Physiol. 1980. 48: 1045-51.
17. SLUTSKY A.S., SHARF S.M., BROWN R., INGRAM R.H.. The effect of oleic acid-induced pulmonary edema on pulmonary and chest wall mechanics in dogs. Am. Rev. Respir. Dis. 1980. 121: 91-6.
18. GRIMBY G., TAKISHIMA T., GRAHAM W., MACKLEM P.T., MEAD J.. Frequency-dependance of flow-resistance in patients with obstructive lung disease. J. Clin. Invest. 1968. 47: 1455-65.
19. CHUNG K.F., KEYES S.J., MORGAN B.M., JONES P.W., SNASHALL P.D.. Mechanisms of airway narrowing in acute pulmonary oedema in dogs: influence of the vagus and lung volume. Clin. Science, 1983. 65: 289-96.
20. HOGG J.C., AGARAWAL J.B., GARDINER A.J.S., PALMER W.H., MACKLEM P.T.. Distribution of airway resistance with developing pulmonary edema in dogs. J. Appl. Physiol.. 1972. 32: 20-4.
21. MICHEL R.P., ZOCCHI L., ROSSI A., CARDINAL G.A., PLOYSONGSANG Y., POULSEN R.S.. MILIC-EMILI J., STAUB N.C.. Does interstitial edema compress airways and arteries? A morphometric study. J. Appl. Physiol., 1987, 62, 1, 108-15.
22. WILSON R.S.. Monitoring the lung: mechanics and volume. Anesthesiology, 1976. 45: 135-45.

A. GRASSINO - C. FRACCHIA - C. RAMPULLA - L. ZOCCHI
Respiratory Muscles in COPD
Bi & Gi Publishers, Verona - Springer-Verlag, London, 1988

The respiratory muscles in acute respiratory failure

R. PARIENTE
Clinical evalutation of respiratory muscle fatigue, Inserm U 226, Hôpital Beaujon, Clichy, France

KEY WORDS: Respiratory muscle; respiratory failure, acute.

Work of breathing depends upon the efficiency of the respiratory muscles and mechanical properties of the lungs. When for a given efficiency of the respiratory muscles the mechanical properties of the lungs are impaired, the load imposed on the respiratory muscles increases. Furthermore, in patients with chronic obstructive pulmonary disease the respiratory muscles, because of hyperinflation, are at a mechanical disadvantage for generating the driving pressure.

These two factors, the increased load to the respiratory muscles and their mechanical disadvantage, may lead in patients with chronic obstructive pulmonary disease to respiratory muscle fatigue.

The detection of this respiratory muscle fatigue is difficult. The available techniques for its detection are the following:
- phrenic stimulation[1]
- changes in the high-to-low ratio of the diaphragmatic EMG[2]
- relaxation rate of the transdiaphragmatic pressure, which is lengthened when the muscle is fatigued[3].

However, all these techniques are difficult to use at the bedside.

Some clinical signs are, however, indicative of respiratory muscle fatigue:
- changes in breathing pattern (rapid and shallow breathing)[4]
- contraction of the accessory muscle and the abdominal muscles.

References

1. AUBIER M., MURCIANO D., LECOCGUIC, VIIRÈS Y.N., PARIENTE R.. Bilateral phrenic stimulation: a simple technique to assess diaphragmatic fatigue in humans. J. Appl. Physiol. 1985, 58, 1, 58-64.
2. GROSS D., GRASSINO A., ROSS W.R.D., MACKLEM P.T.. Electromyogram pattern of diaphragmatic fatigue. J. Appl. Physiol. 1979. 46: 1-7.
3. ESAU S.A., BELLEMARE F., GRASSINO A., PERMUTT S., ROUSSOS C., PARDY R.L.. Changes in relaxation rate with diaphragmatic fatigue in humans. J. Appl. Physiol. 1983, 54: 1353-1360.

4. AUBIER M., MURCIANO D., FOURNIER M., MILIC-EMILI J., PARIENTE R., DERENNE J.Ph.. Central respiratory drive of patients with chronic obstructive pulmonary disease in acute respiratory failure. Am. Rev. Respir. Dis. 1980. 122: 191-200.

Therapeutical Approaches

A. Grassino - C. Fracchia - C. Rampulla - L. Zocchi
Respiratory Muscles in COPD
Bi & Gi Publishers, Verona - Springer-Verlag, London, 1988

Respiratory muscle training in COPD

R. L. Pardy
UBC Pulmonary Research Laboratory St. Paul's Hospital, Vancouver (B.C.), Canada

SUMMARY: Principles used in developing skeletal muscle strength and endurance have for some years been applied to training respiratory muscle performance in normal humans.
In COPD, it is feasible to postulate that increased work of breathing may promote a spontaneous training response; muscular strength is clearly not enhanced.
Training programs have thus been instituted to develop the respiratory muscle function of COPD patients, with particular reference to endurance.
Voluntary isocapnic hyperpnea, inspiratory resistive loading and threshold loading are the main techniques used.
Respiratory muscle endurance in COPD can be improved by selective exercise, though total enhancement of exercise performance has yet to be confirmed.
Intensity, frequency and duration of exercise stimuli have also to be studied further.

KEY WORDS: Respiratory muscle training: COPD, skeletal muscle strength: skeletal muscle endurance; respiratory muscle function: respiratory muscle endurance.

Principles of skeletal muscle training have been applied to the respiratory muscles over the past ten years. To obtain a training response in skeletal muscles, an appropriate training stimulus is applied, which leads to a useful adaptation in the muscle[1, 5]. Three fundamental principles of skeletal muscle training are overload, specificity, and reversibility. The overload principle states that 'for skeletal muscle fibres to increase their structural capability, functional capabilities, or both, the capability must be taxed beyond some critical level'. The specificity principle states that 'the stimulus for adaptation acts on specific skeletal muscle fibres and specific structural and functional organelles within the skeletal muscle fibre that are overloaded beyond the critical threshold for adaptation'. The reversibility principle states that 'the effects of conditioning are transient'.

Based on these principles, conditioning programs are designed to improve muscle function. The conditioning program must contain a training stimulus of sufficient intensity, duration, and frequency, applied over a sufficiently long course to produce a training response. The conditioning program is different for strength training and endurance training[1, 2, 3, 5]. For strength training, low repetitions of a high intensity stimulus are applied (think of the weight lifter). For endurance training, high repetitions of a low intensity stimulus are applied (think of the marathon runner).

Leith and Bradley were the first to systematically apply these principles to the respiratory muscles of normal humans in 1976 . They clearly demonstrated in normal man that respiratory muscle strength increases in response to a specific strength training program and that respiratory muscle endurance increases in response to a specific endurance training program. The cellular changes in the respiratory muscles in response to such specific training have not been studied in man, but it is reasonable to assume that they would be similar to those found in other human skeletal muscles[4], as well as those found in trained respiratory muscles of smaller animals[7, 8]. Strength training results in muscle hypertrophy and endurance training results in increased oxidative capacity, increased blood flow and no change in muscle fibre size. Animal experiments indicate that the respiratory muscles do respond to endurance training stimuli with appropriate cellular adaptations in the muscle fibres[7, 8].

Leith and Bradley's work in normal humans led to a number of studies in a variety of clinical conditions to determine whether similar results could be obtained. The majority of these studies have been undertaken in patients with chronic obstructive pulmonary disease (COPD). In a typical patient with severe COPD, the work of breathing is substantially increased and the respiratory muscles are operating at a length-tension disadvantage caused by hyperinflation[9]. Theoretically, there are two quite different potential consequences of the increased load placed on the respiratory muscles in COPD. At one extreme is the possibility that since the respiratory muscles are chronically working against increased loads, useful adaptations (for example, an endurance training response) may develop. At the other extreme is the possibility that there is a maladaptation in the form of chronic fatigue resulting from chronic overload. Experiments in elastase-induced emphysematous hamsters have shown useful adaptations to occur in the diaphragm. The diaphragm of emphysematous hamsters is less fatiguable than normal and a useful change in the contractile properties of the diaphragm has been observed, whereby its fibres develop maximum force at a shorter length than in control animals[10, 11]. On the other hand, in humans, what evidence is available (and it is not very complete) suggests that useful adaptations to the disease do not occur in the respiratory muscles. Respiratory muscle strength is either normal or decreased[9, 12]. This is not surprising, since the load in COPD is not great enough to evoke a strength training response. Since there are problems comparing respiratory muscle endurance in normal subjects and patients with COPD, it is not possible to state whether endurance is different in these two populations. Recent studies of biopsies of the respiratory muscles in COPD patients submitted to resectional lung surgery have shown that type II fibre atrophy is common[13, 14]. Since similar changes were found in non-respiratory muscles, it can be argued that at least part of these changes may be related to age or associated disease (such as malignancy)[14]. Whatever the explanation, the fact that about half the subjects had atrophy of some respiratory muscle fibres indicates that the disease itself does not have a useful training effect in these individuals. Atrophy implies

Table 1 - Effects of Training on RM Strength in Humans

Subjects	Source	Stimulus	Response (% Increase RM Strenght)
COPD and Tuberculosis	Lecoq 1970 (32)	Max. inspirations	50
	Lecoq 1970	Max. expirations	34
Normal	Leith 1976 (6)	Max. inspirations	55
COPD	Pardy 1981 (18)	Inspiratory loading	Nil
COPD	Reid 1984 (26)	Max. inspirations	53
		Max. inspirations	10

Numbers in brackets indicate reference number in bibliography.

Table 2 - Effects of exercise training on RM endurance in humans

Subjects	Source	Stimulus	Response (% Increase MSV*)
Normal	Robinson & Kjeldgaard (35)	Running	16
Cystic Fibrosis	Keens et al (16)	Swimming/Canoeing	57
Cystic Fibrosis	Orenstein et al (34)	Running	50
COPD	Belman and Kendregan (33)	Arm/Leg Cycling	Nil

* MSV = Maximum sustainable ventilation

Table 3 - Effects of voluntary hyperpnea on RM Endurance

Authors	Subjects	Duration	Frequency	Course	Response (% Increase MSV)	Better than control?
Leith 1976 (6)	4 normals	20-30 min	5/week	5 weeks	19	Yes
Keens 1977 (16)	4 normals	25 min	5/week	4 weeks	22	
Keens 1977 (16)	4 CF	15 min	5/week	4 weeks	55	No
Belman 1980 (17)	10 COPD	15 min	2/day	6 weeks	33	

COPD = chronic obstructive pulmonary disease
CF = cystic fibrosis
MSV = maximum sustainable ventilation (for 15 minutes)

disuse, not overuse. The frequency of its presence in patients with COPD supports the concept that attempts should be made to train their respiratory muscles (and incidentally suggests that resting them could potentially lead to even more atrophy).

Over the last ten years, a fairly substantial literature has accumulated on the effects of respiratory muscle training in patients with COPD[15, 31]. Strength training of the respiratory muscles is relatively less important than endurance training. However, as shown in Table 1, it is encouraging to note that respiratory muscle

Table 4 - Effects of inspiratory Resistive and Threshold (*) Loading on RM Endurance - COPD & CF

Authors	Subjects	Duration	Frequency	Course	Response			Better than
					SIP	MAX R	Endurance vs Set R	control?
Andersen 1977 (15)	10 COPD	30 Min	1/day	8 weeks		+	+	—
Pardy 1981 (18,19)	12 COPD	15 min	2/day	8 weeks		+	+	Yes
Bjerre-Jepson 1981 (20)	14 COPD	15 min	3/day	6 weeks		+		No
Sonne 1982 (21)	6 COPD	30 min	1/day	6 weeks		+		Yes
Asher 1982 (22)	11 CF	15 min	2/day	4 weeks		+		Yes
Chen 1985 (30)	7 COPD	15 min	2/day	4 weeks	+			Yes
(*) Clanton 1985 (29)	4 COPD	7.5 min	3/week	10 weeks	+	+	+	Yes

COPD = chronic obstructive pulmonary disease

CF = cystic fibrosis

SIP = sustainable inspiratory pressure

MAX R= maximum resistance

Endurance vs. Set R = endurance time inspiring against a given resistance or threshold load

\+ = increase

Table 5 - Effects of inspiratory Resistive Loading on RM Endurance - Other Diseases

Authors	Subjects	Duration	Frequency	Course	Response		Better than
					MAX R	Endurance vs Set R	control?
Gross 1980 (36)	6 Quad	15 min	12/week	8 weeks	+	+	
Di Marco 1982 (37)	10 MD	15 min	2/day	6 weeks	+	+	
Martin 1983 (38)	1 Myopathy	15 min	2/day	15 weeks	+		

Quad = quadriplegia
MD = muscular dystrophy

strength can be increased in both normal and diseased humans[6, 18, 26, 32]. Potentially, therefore, individuals with weak respiratory muscles may show a useful adaptation if exposed to the appropriate strength training conditioning program.

Endurance training of the respiratory muscles can be achieved by both specific and non-specific conditioning programs. The obvious non-specific program is total body exercise. If exercise is of sufficient intensity and duration, minute ventilation may be increased to a sufficient degree to have an endurance training effect on the respiratory muscles of both normal and diseased humans (Tab. 2)[16, 33-35]. Patients with severe lung diseases are unlikely to obtain such a non-specific respiratory muscle training effect, since they are unable to exercise at a sufficient intensity or for long enough to produce any useful adaptation in their respiratory muscles. Specific respiratory muscle endurance training programs have therefore been used in such patients.

Specific respiratory muscle endurance training programs applied to both normal and diseased humans are of three basic types. The first is voluntary isocapnic hyperpnea[6]. In this form of training, individuals maintain high target levels of ventilation over periods up to 15 minutes. The response to such training is measured as a change in the maximum sustainable ventilation (MSV). The second is inspiratory resistive loading, in which resistive loads are applied to the inspiratory muscles over five to fifteen minutes[15, 18]. Several simple devices employing this approach are now available commercially. The response to such training is measured either as an increase in the maximum tolerable resistance over a specified period of time or an increase in the time a given load is tolerated. The third form of specific respiratory muscle endurance training is threshold loading[29]. With this form of training, the inspiratory pressure load is determined by the amount of weight applied to the inspiratory port of a one-way valve. The response is assessed by a change in the time a subject can breathe against a given load.

Voluntary hyperpnea (Tab. 3)[17], inspiratory resistive loading (Tab. 4)[15, 18-21, 23-28, 30, 31] and inspiratory threshold loading (Tab. 4)[29] may lead to improved respiratory muscle endurance in patients with obstructive lung disease. As shown in Table 5, inspiratory resistive loading has also led to improved endurance in small numbers of patients with several different chest wall diseases.

A number of comments can be made from a closer look at the studies listed in Tables 3 and 4. First, only small numbers of patients were present in each study. Second, the duration of training sessions has been relatively short (in the order of 7 - 30 minutes). Third, although most endurance athletes only do intense training on alternate days, there is only one study of alternate day training of the respiratory muscles in this list[29]. In most studies daily training was performed. Fourth, the course of training has been short in all studies. Fifth, quite marked variability in the response was observed. Several studies[20, 24], not included in Tables 3 and 4, showed no response to training, while an improvement in respiratory muscle endurance was found in those listed. The degree of response was variable and possibly reflects the intensity of the training program used. Sixth, some of the studies have not included control groups. However, controls have been included in enough studies to conclude that specific training is associated with improved respiratory muscle endurance in some COPD patients.

Two other observations (not included in the tables) have been made in patients with obstructive lung diseases after specific respiratory muscle endurance training. First, several studies have shown an increase in strength in a variety of diseases including COPD[22, 24, 26, 30], while other studies showed no such effect[18]. It appears that if the intensity of the training is sufficient (i.e. the resistive load is large enough), a strength training response may occur.

Second, increased exercise tolerance has been shown in a number of studies of patients with COPD[17, 18, 21] or cystic fibrosis[16, 22]. No change in exercise tolerance was found in another two studies of patients with COPD[24, 30] and, in a further study, specific respiratory muscle training was no better than a total body exercise training program in patients with cystic fibrosis[16]. It can be concluded at this point in time that it is possible to improve respiratory muscle endurance with specific training in patients with obstructive lung disease. Whether an improvement in exercise performance will result from such training is still open to debate. Our own work suggests that only some patients will improve their exercise performance. They are identified as subjects with evidence of inspiratory muscle fatigue contributing to exercise limitation in pre-training exercise tests[18].

Despite my conclusion that it is possible to train the respiratory muscles of patients with COPD, a number of unanswered questions remain. First, what constitutes the ideal training program cannot be determined at this time, since we lack controlled studies comparing conditioning programs containing groups exposed to differing intensities, frequencies and durations of appropriate training stimuli. Second, no one has tried decreasing the frequency of training sessions once a training response has been achieved to determine what is the minimum amount of training that is necessary to maintain any improvement that has been achieved. Skeletal muscle literature would suggest that a maintenance program can be at an appreciably lower level than the initial training program. Third, studies of the long-term effects of respiratory muscle training on symptoms, morbidity and mortality, while absent from literature and difficult to do, are important, since this form of therapy has become popular in rehabilitation programs. We need to know whether the increasing effort being expended in this direction is justified by a lessening morbidity, improved sense of well-being and/or decreased mortality. Finally, it has recently been shown that malnourished individuals have malnourished and weak respiratory muscles[39, 41]. Since malnutrition is at least not uncommon in the later stages of obstructive lung disease, we need to give some thought to the concept of improving the nutritional status of patients either before attempting respiratory muscle training or in association with it.

References

1. FAULKNER J.A.: New perspectives in training for maximum performance. JAMA 1968. 205: 741-746.
2. ANDERSON T., KERNEY J.T.: Effects of three resistance training programs on muscular strength and absolute and relative endurance. Res. Quart. Exerc. Sport. 1982. 53: 1-7.
3. DELATEUR B.J., LEHMANN J.F., GIANCONI R.: Mechanical work and fatigue: Their role in the development of muscle work capacity. Arch. Phys. Med. Rehab. 1976. 57: 319-324.
4. HOLLOSZY J.O.: Adaptations of muscular tissue to training. Prog. Cardiovasc. Dis. 1976. 18: 445-458.
5. KATCH F.I., FREEDSON P.S.: Effects of different modes of strength training in body composition and anthropometry. Clinics in Sports Med. 1986. 5: 413-460.

6. LEITH D.E., BRADLEY M.: Ventilatory muscle strength and endurance training. J. Appl. Physiol. 1976. 41-508-516.
7. KEENS T.G., CHEN V., PATEL P., O'BRIEN P., LEVISON H., IANUZZO C.D.: Cellular adaptations of the ventilatory muscles to a chronic increased respiratory load. J. Appl. Physiol. 1978. 44: 905-908.
8. LIEBERMAN D.A., MAXWELL L.C., FAULKNER J.A.: Adaptation of guinea pig diaphragm muscle to aging and endurance training. Am. J. Physiol. 1972. 222: 556-560.
9. ROCHESTER D.F., ARORA N.S., BRAUN N.M.T., GOLDBERG S.K.: The respiratory muscles in chronic obstructive pulmonary disease (COPD). Bull Europ Physiopathol Respir. 1979. 15: 951-975.
10. FARKAS G.A., ROUSSOS C.: Adaptability of the hamster diaphragm to exercise and/or emphysema. J. Appl. Physiol. 1982. 53: 1263-1272.
11. SUPINSKI G.S., KELSEN S.G.: Effect of elastase-induced emphysema on the force generating ability of the diaphragm. J. Clin. Invest. 1982. 70: 978-988.
12. ROCHESTER D.F., BRAUN N.M.T., ARORA N.S.: Respiratory muscle strength in chronic obstructive pulmonary disease. Am. Rev. Respir. Dis. 1979. 119: 151-154.
13. SANCHEZ J., DERENNE J.P., DEBESSE B., RIQUET M., MONOD H.. Typology of the respiratory muscles in normal men and in patients with moderate chronic respiratory disease. Bull. Europ. Physiopath. Respir. 1982. 18: 901-914.
14. HUGHES R.L., KATZ H., SAHGAL V., CAMPBELL J.A., HARTZ R., SHIELDS T.W.: Fiber size and energy metabolites in five separate muscles from patients with chronic obstructive lung disease. Respiration. 1984. 44: 321-328.
15. ANDERSEN J.B., DRAGSTED L., KANN T., JOHNASEN T., NIELSEN K.B., KARBO E., BENTZEN L.: Resistive breathing training in severe chronic obstructive pulmonary disease. A pilot study. Scan. J. Respir. Dis. 1977. 60: 151-155.
16. KEENS T.G., KRASTINS I.R.B., WANNAMAKER E.M., LEVISON H., CROZIER D.N., BRYAN A.C.: Ventilatory muscle endurance training in normal subjects and patients with cystic fibrosis. Am. Rev. Respir. Dis. 1977. 116: 853-860.
17. BELMAN M.J., MITTMAN C.: Ventilatory muscle training improves exercise capacity in chronic obstructive pulmonary disease patients. Am. Rev. Respir. Dis. 1980. 121:273-280.
18. PARDY R.L., RIVINGTON R.N., DESPAS P.J., MACKLEM P.T.: The effects of inspiratory muscle training on exercise performance in chronic airflow limitation. Am. Rev. Respir. Dis. 1981. 123: 426-433.
19. PARDY R.L., RIVINGTON R.N., DESPAS P.J., MACKLEM P.T.: Inspiratory muscle training compared with physiotherapy in patients with chronic airflow limitation. Am. Rev. Respir. Dis. 1981. 123: 421-425.
20. BJERRE-JEPSON K., SECHER N.H., KOK-JENSEN A.: Inspiratory resistance training in severe chronic obstructive pulmonary disease. Eur. J. Respir. Dis. 1981. 62: 405-411.
21. SONNE C.J., DAVIS J.A.: Increased exercise performance in patients with severe COPD following inspiratory resistive training. Chest 1982. 81: 436-439.
22. ASHER M.I., PARDY R.L., COATES A.L., THOMAS E., MACKLEM P.T.: The effects of inspiratory muscle training in patients with cystic fibrosis. Am. Rev. Respir. Dis. 1982. 126: 855-859.
23. MORENO R., MORENO R., GIUGLIANO C., LISBOA C.: Entrenamiento muscular inspiratorio en pacientes con limitacion cronica del flujo aereo. Rev. Med. Chile. 1983. 111: 647-653.
24. AMBROSINO N., PAGGIARO P.L., ROSELLI M.G., CONTINI V.: Failure of resistive breathing training to improve pulmonary function tests in patients with chronic obstructive pulmonary disease. Respiration. 1984. 45: 455-459.
25. LARSON M., KIM M.J.: Respiratory muscle training with the incentive spirometer resistive breathing device. Heart and Lung. 1984. 13: 341-345.
26. REID W.D., WARREN C.P.W.: Ventilatory muscle strength and endurance training in elderly subjects and patients with chronic airflow limitation: A pilot study. Physiol. Canada. 1984. 36: 305-311.

27. JEDERLINC P.J., MUSPRATT J.A., MILLER M.J.: Inspiratory muscle training in clinical practice. Chest. 1984. 86: 870-873.
28. ANDERSEN J.B., FALK P.: Clinical experience with inspiratory resistive breathing training. Int. Rehabil. Med. 1984. 6: 183-185.
29. CLANTON T.L., DIXON G., DRAKE J., GADEK J.E.: Inspiratory muscle conditioning using a threshold loading device. Chest. 1985. 87: 62-66.
30. CHEN H., DUKES R., MARTIN B.J.: Inspiratory muscle training in patients with chronic obstructive pulmonary disease. Am. Rev. Respir. Dis. 1985. 131: 251-255.
31. ALDRICH T.K., KARPEL J.P.: Inspiratory muscle resistive training in respiratory failure. Am. Rev. Respir. Dis. 1985. 131: 461-462.
32. LECOQ A., DELHEZ L., JANSSENS F., PETIT J.M.: Réentrainment de la fonction matrice ventilatoire chez des insuffisants respiratoires chroniques. Acta. Tub. et Pneumo. Belg. 1970.
33. BELMAN H.J., KENDERGAN B.A.: Exercise training fails to increase skeletal muscle enzymes in patients with chronic obstructive disease. Am. Rev. Respir. Dis. 1981. 123: 256-261.
34. ORENSTEIN D.M., FRANKLIN B.A., DOERSHUK C.F., HELLERSTEIN H.K., GERMANN K.J., HOROWITZ J.G., STEIN R.C.: Exercise conditioning and cardiopulmonary fitness in cystic fibrosis. Chest. 1981. 80 392-398.
35. ROBINSON E.P., KJELDGAARD J.M.: Improvement in ventilatory muscle function with running. J. Appl. Physiol. 1982. 52: 1400-1406.
36. GROSS D., LADD H.W., RILEY E.J., MACKLEM P.T., GRASSINO A.: The effect of training on strength and endurance of the diaphragm in quadriplegia. Am. J. Med. 1980. 68: 27-35.
37. DIMARCO A.F., KELLING J., SAJOVIC M., JACOBS I., SHIELDS R., ALTOSE M.D. Respiratory muscle training in muscular dystrophy (abstract). Clin. Res. 1982. 30: 427A.
38. MARTIN R.J., SUFIT R.L., SINGEL S.P., HUDGEL D.W., HILL P.L.: Respiratory improvement by muscle training in adult-onset acid maltase deficiency. Muscle Nerve. 1983. 6: 201-203.
39. ARORA N.S., ROCHESTER D.F.: Effect of body weight and muscularity on human diaphragm muscle mass, thickness and area. J. Appl. Physiol. 1982. 52: 64-70.
40. ARORA N.S., ROCHESTER D.F.: Respiratory muscle strength and maximal voluntary ventilation in undernourished patients. Am. Rev. Respir. Dis. 1982. 126: 5-8.
41. ROCHESTER D.F., BRAUN N.M.T.: Determinants of maximal inspiratory pressure in chronic obstructive pulmonary disease. Am. Rev. Respir. Dis. 1985. 132: 42-47.

A. GRASSINO - C. FRACCHIA - C. RAMPULLA - L. ZOCCHI
Respiratory Muscles in COPD
Bi & Gi Publishers, Verona - Springer-Verlag, London, 1987.

Treatment of patients with respiratory failure during wakefulness and sleep: use of tank ventilator

M. SCHIAVINA
S. Orsola Hospital, Bologna, Italy

SUMMARY: The tank ventilator, a long established method of treating acute and chronic respiratory failure, including exacerbation of C.O.P.D., is the object of increasing interest. While the functional principle has remained unchanged from the early models, modern ventilators allow a more sensitive and systematic control of respiratory variables, with both positive and negative pressure settings.

Tank-assisted respiration can provide an alternative to intubation in a variety of respiratory conditions, including kyphoscoliotic and neuromyopathic forms.

Practical examples are provided.

The need for more satisfactory monitoring to supplement arterial blood gas analysis of tank-ventilated patients is stressed.

KEY WORDS: Respiratory failure; wakefulness; sleep; tank ventilation; COPD; tank-assisted respiration; intubation; kyphoscoliosis; neuromyopathies.

Against a background of growing interest in the use of tank ventilators, it is appropriate to attempt a succint definition of their indications, the principles on which they work and the factors indispensable to their efficacy.

The original aim of the tank ventilator was to treat acute and chronic respiratory failure. Artificial ventilation by means of an extrathoracic lung is not a new technique. The first iron lung was described by Alfred Jones of Lexington, Kentucky, in 1864. It was in 1876, however, that the true forerunner of today's tank respirators was constructed in Paris by Woillez. By 1928, Drinker was able to introduce into clinical practice a considerably improved model, which proved the salvation of many lives during a poliomyelitis epidemic.

Functionally, the principle on which the tank respirator is based is the production of sinusoid pressure oscillations inside the tank by means of a large motorized bellows or piston. Once the patient's body is placed inside the respirator, the negative pressure created within the tank determines decreased intrathoracic pressure, which is compensated by the arrival of air entering through the nose and mouth

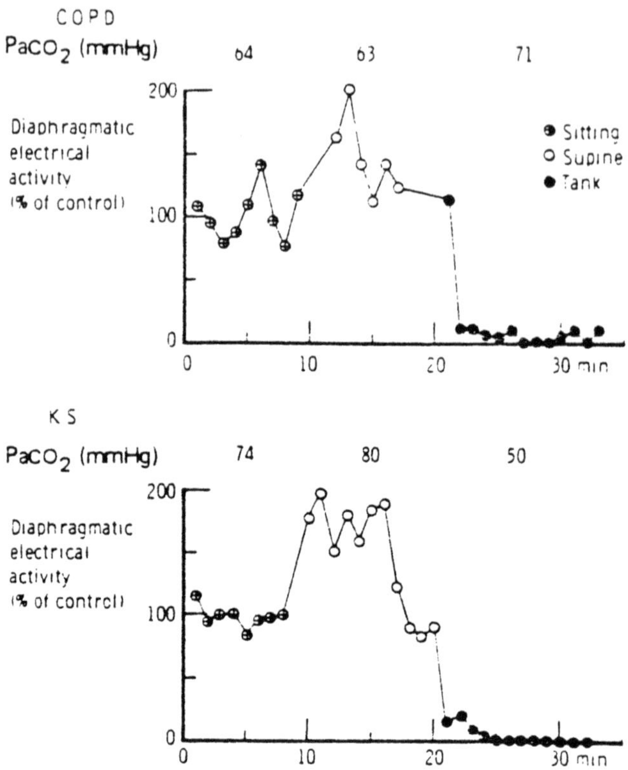

Fig. 1 - The electrical activity of the diaphragm.

and passing into the lungs in response to the pressure differential between the buccal cavity and the alveoli. By the same token, expiration occurs when positive pressure is created inside the tank, determining increased intrathoracic pressure and reversing the pressure differential between the buccal cavity and the alveoli. With pressure within the lungs higher than that outside, expiration is the logical consequence. The tank respirator thus sets up a ventilation mechanism close to that of natural breathing.[2,4-5]

While Drinker's system allowed for an inspiration/expiration ratio of 1, with no regulation of ventilatory pauses at the end of inspiration and expiration, the system adopted by us is based on the use of Pulmolife tank respirator, by means of which the following variables can be controlled:

1) regulation of positive and negative pressure up to values of 60 cm H_2O, with independent setting and re-setting;
2) independent variation of the duration of each inspiration and expiration within a range of 0.83 - 2.8 seconds, allowing for adjustment of the inspiration/expiration ratio;

3) possible pauses between the end of each expiration and the beginning of each inspiration, and vice versa;
4) breathing rate variable from 14 to 24 cycles per minute.

Though the body respirator has featured in clinical practice for about 60 years, its true value in the treatment of respiratory failure has never been satisfactorily catalogued. There have been documented case histories illustrating its use and coloured by a degree of scepticism with regard to its usefulness. This attitude undoubtedly stems from the ad-hoc approach that was for a long time prevalent when tank respirators were used and studied. It was not until 1977 that Rochester produced a more comprehensive study of the effects of tank-assisted respiration.[16] There had previously been no satisfactory account of the subject.

Many patients with chronic hypercapnia related to chronic obstructive pulmonary disease, constrictive diseases of the thorax or neuromuscular disorders present a marked increase of respiratory effort, with a high degree of fatigue in the respiratory muscles.

The tank respirator can, in such patients, take over almost completely the function of the respiratory muscles in the presence of an excessive work load.

Rochester greatly furthered the study of the tank respirator's effects on ventilation and the electrical activity of the respiratory muscles. He formulated a number of important conclusions (Fig. 1).

1) The electrical activity of the diaphragm and the inspiratory muscles is appreciably reduced or disappears in C.O.P.D. and kyphoscoliotic patients.[1,8,10] The loss of this muscular activity is readily and immediately seen if a negative inspiratory pressure of –20, –25 cm. H_2O is set up around the body.
2) If, in the absence of activity of the inspiratory muscles, the negative pressure is reduced to –15, –10 cm. H_2O, the muscles are reactivated.
3) The reduction or disappearance of inspiratory muscular activity is accompanied by a marked reduction of dyspnea.
4) The mechanism by which the tank respirator rapidly inhibits the muscular activity attendant on inspiration is not clear. It is, however, not related to a fall of $PaCO_2$, which may occasionally rise rather than decrease.

Rochester's observations thus suggest that the weakness and fatigue of the inspiratory muscles reduce ventilatory capacity, with consequent hypercapnic respiratory failure. Resting the respiratory muscles by means of the tank respirator thus not only prevents muscular fatigue, but also promotes enhanced muscular efficiency and ventilatory capacity.

On the basis of this premise and the principle that rest is the most appropriate treatment for muscular fatigue, the tank ventilator finds its principal application in the treatment of acute and chronic respiratory failure related to chronic respiratory failure related to chronic obstructive bronchopulmonary disease, alterations of the thoracic cage, neuromuscular disorders or disturbed breathing, as well as in the weaning of intubated patients (Tab. 1)[3].

Tab. 1 - Indications for use of tank respirator

Acute respiratory failure (psychotropic drugs, morphine ...).
Relapses in chronic bronchopulmonary disease (excluding asthma), weaning of intubated patients.
Alterations of thoracic cage (kyphoscoliosis ...).
Neuromuscular diseases (Duchenne, Dubowitz, poliomyelitis).
Breathing disorders during sleep.

Our 15 years' experience of using the tank respirator with simultaneous monitoring of arterial blood gas levels prompts a number of important conclusions and observations.

The first and most frequent indication for the use of the tank respirator is to provide an alternative to nasotracheal intubation in the management of respiratory failure, both acute and especially chronic, including relapsing C.O.P.D. – related forms. Indications for the use of a tank ventilator in such cases are intense dyspnea, hypercapnia and hypoxemia. Tank-assisted respiration reduces hypercapnia, while oxygen therapy can be administered without increasing Pa CO_2 and pulmonary hypertension can be relieved. The most significant clinical sign of the resting of the respiratory muscles is the almost immediate reduction of dyspnea in such cases. The capacity of the tank respirator to improve ventilation in patients presenting relapses of stable C.O.P.D. related to infective or metabolic complications is, in fact, an extremely important feature of such therapy, providing an alternative to nasotracheal intubation. Since there are various contraindications to intubation in hyperinsufflated patients presenting uncleared infections, recourse to tracheostomy is often necessary. Tank therapy of the acute phases of the disease often allows respiratory failure to be managed without the risk of secondary infections or tracheal stenosis attendant on intubation. The treatment also allows for effective weaning of intubated patiens who would otherwise experience difficulty in acquiring independence from tube ventilation.

Two problems arise when C.O.P.D. patients with respiratory failure are treated by means of a tank respirator. The first is that the reduction of hypercapnia is not always rapid. The second is that some patients develop an obstruction of the upper airways.

The blood gas profile sometimes presents a worsening of $PaCO_2$, which can be directly related to the formation of an obstruction in the upper airways, determined either by uncoordinated activation of the relevant muscles or by the failure of the vocal cords to open during machine-induced inspiration. In both cases, the cause seems to be a lack of synchronization of the vocal cords and upper airway muscles to the breathing cycle determined by the tank respirator.[6,7]

This phenomenon frequently takes the form of obstructive sleep apnea. As such, it has been observed even in normal subjects undergoing tank-assisted ventilation during sleep. Apnea of this type, as observed by P. Macklem, may be related to the non-activation of the upper airway muscles, in turn caused by the cut-out of diaphragm activity in the presence of extrathoracic negative pressure.

The tank ventilator is referred to in literature as the negative-pressure ventilator, as the principle of the device should provide for application of negative pressure as the sole stimulus to efficient inspiration and expansion of the thoracic cage. Expiration should be passively induced when pressure within the tank respirator is once more atmospheric. In practice, we constantly apply positive expiratory pressure. Blood gas analysis, confirmed by patients' impressions, suggests that positive pressure is a necessary prerequisite for more efficient expiration (Tab. 2).

Tab. 2 - Blood gas analysis

	Base	Pi-45/PeO	Pi-45/Pe+20
PaO_2	45	62	68
$PaCO_2$	77	71	52
pH	7.28	7.29	7.3 4

Incorrect application of positive pressure can, however, prejudice any improvement of $PaCO_2$ or even cause a worsening of the parameter, probably following a collapse of the airways during expiration.

Another important indication for the use of the tank respirator is for treatment of respiratory failure related to diseases of the thoracic cage, particularly kyphoscoliosis. The normal evolution for patients presenting such diseases, with consequently reduced thoracic compliance and increased respiratory effort that leads in turn to muscular fatigue, is the onset of alveolar hypoventilation. Its consequences are hypercapnia, hypoxemia and cor pulmonale. The condition is often irreversible, with an understandably poor prognosis. As in C.O.P.D. sufferers, intermittent positive-pressure ventilation determines a temporary increase of pulmonary compliance and PaO_2, but cannot be indicated for chronic use. In such patients, the tank respirator can obviate the need for permanent tracheostomy.

Tank respirator treatment of kyphoscoliosis-related respiratory failure, while determining an appreciable improvement of the arterial blood gas profile in the acute phase (Tab. 3), poses the problem of long-term and nocturnal treatment, with occurrence of mainly central hypopnea and apnea and oxygen shortage during REM sleep. In patients presenting severe hyperventilation when awake and

Tab. 3- Tank respirator treatment of kyphoscoliosis-nocturnal related respiratory failure.

KyPHO	M.G.		T.G.		C.G.		F.D.	
	Base	Pi-30/Pe+25	Base	Pi-21/Pe+14	Base	Pi-35/Pe+30	Base	Pi-40/Pe+35
PaO_2	44	62	42	68	40	54	36	44
$PaCO_2$	59	40	54	46	66	54	76	52
pH	7.36	7.50	7.38	7.38	7.37	7.46	7.38	7.39
SaO_2	78	93	77	92	75	86	69	80

oxygen desaturation during sleep, the alternative to tank ventilator treatment is the use of a cuirass ventilator at home. The use of cuirass-assisted ventilation is controversial, its effects inferior to those obtained with the tank respirator. A study by Guilleminault of cuirass ventilation in 4 severe kyphoscoliotic patients showed a worsening of apnea during sleep, with no corresponding improvement in ventilation. It should, however, be noted that commercially available cuirasses (Monaghan or body shell) are not easily used in kyphoscoliotic patients because of their muscular deformity. For this reason, as an alternative to the iron lung we are evaluating the possibilities of using a purpose-measured polyethylene cuirass which could be readily used at the patient's home to provide an efficient ventilatory aid. The problem is still at the study stage.

Another indication for the tank respirator is in patients suffering from neuromuscular diseases, in which marked reduction of inspiratory muscle activity necessitates chronic ventilatory treatment. This is necessary not only for improved ventilation and quality of life, but also in terms of life expectancy. The normal treatment in such cases is tracheostomy. Severe oxygen desaturation during sleep, associated with or aggravated by mainly central hypopnea or apnea, has also been observed in myotonic dystrophy (Steirnert), progressive muscular dystrophy and rigid spine syndrome (Dunowitz). The tank respirator provides an efficacious first-choice treatment in such cases.

The case history of an 11-year-old boy with rigid spine syndrome shows improvement in response to tank respirator treatment, with an appreciable bettering of the arterial blood gas profile a year later (Tab. 4). Polysomnographic study of the patient showed a severe fall-off of oxygen saturation at the onset of sleep. Breathing became polypnoic, with intercostal muscular activity showing increased amplitude. Sporadic occurrences of central apnea led to a further rapid reduction of SaO_2. Following such attacks, SaO_2 would rise again. These continuous apnea-related oscillations of the parameter during the course of the night led to its progressive reduction, reaching mean values of 40-50% after approximately 3 hours. Tank respirator treatment determined appreciably improved SaO_2, mainly by virtue of the elimination of central apnea and hypopnea. Despite spasms of obstructive apnea lasting 10-20 seconds and following each other in rapid succession, SaO_2 was maintained at values not lower than 70%.[9,11,15]

Tab. 4- The case history of an 11 year old boy.

	Base	Pi-45/Pe+35	Pi-45/Pe+35 O_2 1.5 l/min.	Base
PaO_2	47	58	97	67
$PaCO_2$	75	53	48	42
pH	7.24	7.36	7.33	7.38
SaO_2	74	88	96	92
	26.7.85	*27.7.85*	*6.5.86*	*8.8.86*

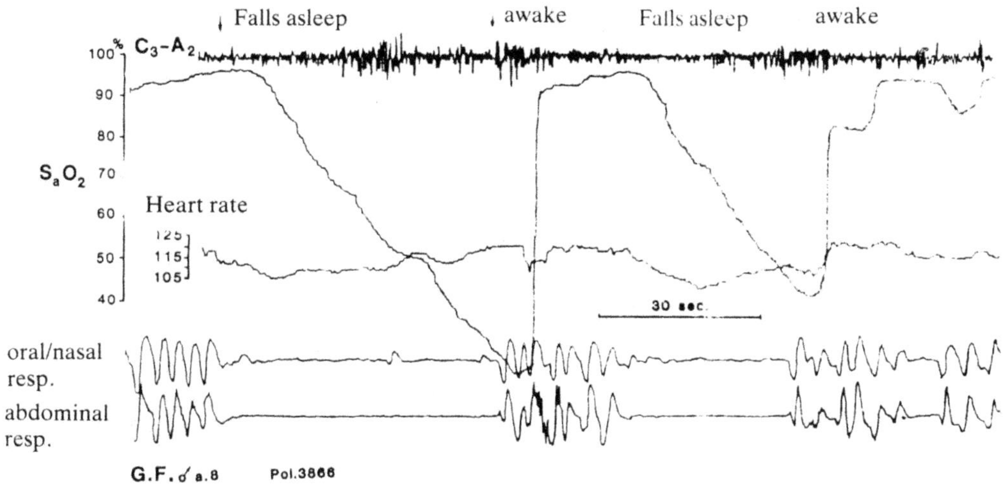

Fig. 2 - Sleep registration of a case of central alveolar hypoventilation

	PaO_2	$PaCO_2$	pH	SaO_2
At time of admission	58	56	7.28	88.4
Spontaneous hyperventilation	81	28	7.46	95.2
0_2 3 l/min	61	75	7.11	89.6
Tank ventilator Pi 30 – Pe 30	90	42	7.28	96.2
Tank ventilator Pi 15 – Pe 15	60	70	7.22	89.2

Tab. 5 - A case of central alveolar hypoventilation with respiratory failure during sleep.

While patients presenting central alveolar hypoventilation do not suffer from weakened inspiratory muscles or increased ventilatory work-load, tank respirator treatment can provide an alternative to intubation, the rocking bed or the diaphragm pace-maker. Examination of a case of central alveolar hypoventilation with respiratory failure during sleep shows that the onset of sleep is accompanied by occurrences of central apnea which reduce SaO_2 levels to values under 40%. The patient in question, an 8-year-old boy, had developed these symptoms following a viral encephalitis of the encephalic trunk and hypothalamus. Apnea lasted until the patient awoke and started again as soon as he fell asleep once more (Fig. 2). Polysomnographic study did not show obstructive apnea. During a second recording, O_2 administration at a flow rate of 3 l/min led to modest CO_2 retention while the patient was awake. The retention became severe during sleep, with $PaCO_2$ rising to 75 mmHg (Tab. 5). Treatment of such a case only by O_2 administration during sleep would obviously have been unacceptable.[13-14,20]

The patient was therefore ventilated by a tank respirator with an inspiratory

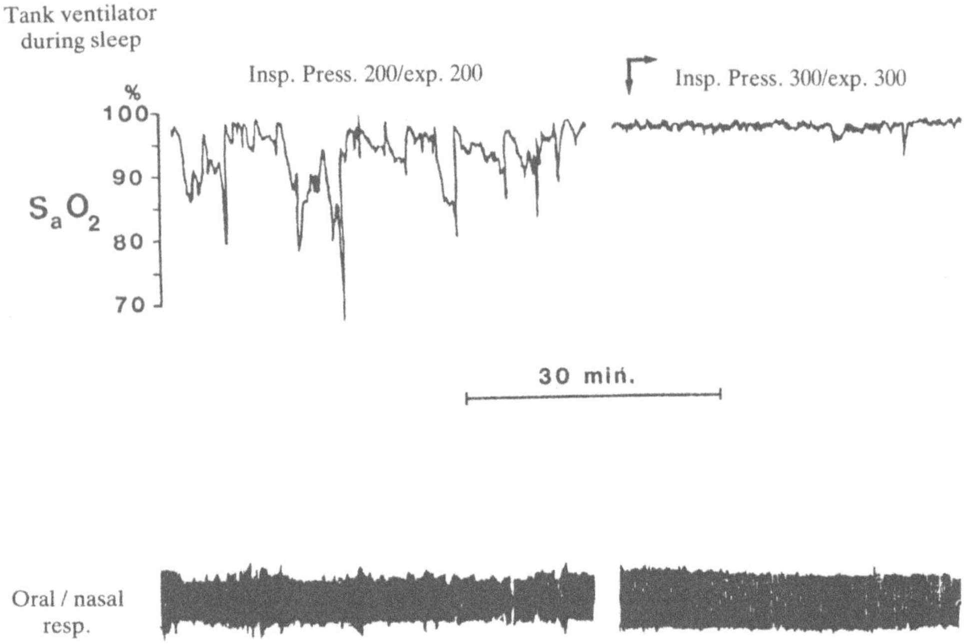

Fig. 3 - Patient was ventilated by a tank respirator.

pressure of –20 cm. H_2O and an expiratory pressure of + 20 cm. H_2O, at a ventilatory frequency of 20 cycles/min. and an O_2 flow rate of 2 l/min. (Fig. 3). In response to this ventilatory treatment, the patient's SaO_2 values during sleep oscillated between 95% and 70%. With inspiratory and expiratory pressures af 30 cm. H_2O, SaO_2 remained at 90-95%, while below these pressures hypoventilation immediately set in, with a consequent fall of SaO_2 proportional to pressure reduction. Reducing the tank ventilator inspiratory and expiratory pressures by 15 cm. H_2O induced cyanosis and an altered blood gas profile during sleep, with PaO_2, $PaCO_2$ and pH showing values of 60 mm Hg, 70 mmHg and 7.11 U respectively. With resetting of pressure levels to 30 cm. H_2O, these values were modified to 90 mmHg, 42 mmHg and 7.28 U respectively.

This finding bears out Drinker's affirmation that there is an (inspiratory) pressure threshold below which the tank respirator does not increase minute ventilation. Above this pressure threshold, the parameter increases in proportion to increases of the tank respirator pressure setting. The main problems in use of the tank respirator are thus not only in accustoming the patient, but also in determining the pressure settings, the duration of each session, blood gas monitoring and the use of oxygen.[12,18]

We have seen that the tank respirator rapidly reduces or abolishes electrical activity of the diaphragm and inspiratory muscles, as a result of which tank ventilator therapy is particularly appropriate for cases of muscular fatigue. While the theoretical basis of tank respirator treatment is the application of negative pressure to the entire body, in practice positive pressure is also applied, determining readier patient compliance and enhanced ventilation and blood gas profile. The advisability of a negative pressure setting higher or lower than the positive pressure level should be individually evaluated for each case on its merits.

With regard to the duration of each session in the respirator, the recommended routine is 3-4 3-hour sessions per day for patients presenting chronic respiratory failure. This also applies to those in whom acute-phase respiratory failure has been successfully cleared. When SaO_2 shows an appreciable drop during sleep, as seen especially in neuromuscular diseases, the patient should remain all right in the tank respirator.[17,19]

Monitoring of the patient while in the tank respirator is performed only by arterial blood gas analysis. In so routine a therapy, this is a shortcoming.

In our opinion, as well as an electromyographic study of the respiratory muscles, plethysmograph ventilatory monitoring, continuous SaO_2 monitoring and transcutaneous electrode readings of blood gas should be performed. This would allow for a more correct and accurate evaluation of the effects of tank ventilation on individual patients during wakefulness and sleep.

References

1. BERGOFSKY E.H., TURINO G.M., FISHMAN A.P., Cardiorespiratory failure in Kyphoscoliosis, Medicine 1959, 38: 263-317.
2. BRAUN N.M., ARORA N.S., ROCHESTER D.F. Respiratory muscle and pulmonary function in polymyositis and other proximal myopathies. Thorax 1983, 38: 616-623.
3. CARDUS D., VALLBONA C., SPENCER W.A., Effects of three kinds of artificial respirators on the pulmonary ventilation and arterial blood of patients with chronic respiratory insufficiency. Dis. Chest. 1966, 50: 297-306.
4. CHERNIAK R.M., SVANHILL E. Long-term use of intermittent positive-pressure breathing (IPPB) in chronic obstructive pulmonary disease. Am. Rev. Respir. Dis. 1976, 113: 721-728.
5. COLLIER C.R., AFFELDT J.E., Ventilatory efficiency of the cuirass respirator in totally paralyzed chronic poliomyelitis patients. J. Appl. Physiol. 1954, 6: 531-538.
6. CURRAN, F.J. Night ventilation by body respirators for patients in chronic respiratory failure due to late stage Duchenne muscular dystrophy. Arch. Phys. Med. Rehab. 1981, 62: 270-274.
7. FISCHER D.A., PRENTICE W.S., Feasibility of home care for certain respiratory-dependent restrictive or obstructive lung diseases in patients. Chest 1982, 82: 739-743.
8. FULKERSON W.J., WILKINS J.K., ESBENSHADE A.M., ESKIND J.B., NEWMAN J.H. Life threatening hypoventilation in Kyphoscoliosis: successful treatment with a molded body brace-ventilator. Am. Rev. Respir. Dis. 1984, 129: 185-187.
9. GUNELLA G., CARA M. La insufficienza respiratoria. Vol. III Bologna Aulo Gaggi, 1973.

10. GUILLEMINAULT C., KURLAND G., WINKLE R., MILES L.E. Severe kyphoscoliosis, breathing and sleep. Chest, 1981, 79: 6-626.
11. HYLAND R.H., HYTCHEON M.A., PERL A., BOWES G., ANTHONISEN N.R., ZAMEL N., PHILIPSON E.A. Upper airway occlusion induced by diaphragm pacing for primary alveolar hypoventilation: Implications for the pathogenesis of obstructive sleep apnea. Am. Rev. Respir. Dis. 1981, 124: 180-185.
12. MARINO W., BRAUN N.M.T. Reversal of the clinical sequelae of respiratory muscle fatigue by intermittent mechanical ventilation. (Abstract). Am. Rev. Respir. Dis. 1982, 125 (Part 2): 85.
13. McCLEMENT J.H., CHRISTIANSON L.C., HUBAYTAR R.T., SIMPSON D.G. The body-type respirator in the treatment of chronic obstructive pulmonary disease. Ann. N.Y. Acad. Sci. 1965, 121: 746-750.
14. MEZON B.L., WEST P., ISRAELS J., KRYGER M. Sleep breathing abnormalities in kyphoscoliosis. Am. Rev. Respir. Dis. 1980, 122: 617.
15. PARDY R.L., RIVINGTON R.N., DESPAS P.J., MACKLEM P.T. The effects of inspiratory muscle training on exercise performance in chronic airflow limitation. Am. Rev. Respir. Dis. 1981, 123; 426-433.
16. ROCHESTER D.F., BRAUN N.M.T., LAINE S. Diaphragmatic energy expenditure in chronic respiratory failure: The effect of assisted ventilation with body respirators. Am. J. Med. 1977, 63: 223-231.
17. SCHARF S.M., FELDMAN N.T., GOLDMAN M.D., INGRAM R.H. Mechanism producing inspiratory upper airway obstruction during electrophrenic respiration and ventilation with an iron lung (Abstract). Am. Rev. Respir. Dis. 1977, 115 (Part 2): 160.
18. SEMB C., ERIKSON H., REFSUM H.E. Ten years experience with tank respirators in the treatment of post-operative respiratory failure. Acta Chir. Scand. 1961, 283 (Suppl): 39-44.
19. SPLAINGARD M.L., JEFFERSON L.S., HARRISON G.M., Survival of patients with respiratory insufficiency secondary to neuromuscular disease treated at home with negative pressure ventilation (NPV). (Abstract). Am. Rev. respir. Dis. 1982, 125: 139.
20. WIERS P.W.J., LeCOULTRE R., DALLINGA O.T., VAN DIJIL W., MEINESZ A.F., SLUITER H.J. Cuirass respirator treatment of chronic respiratory failure in scoliotic patients. Thorax 1977, 32: 221-228.

A. GRASSINO - C. FRACCHIA - C. RAMPULLA - L. ZOCCHI
Respiratory Muscles in COPD
Bi & Gi Publishers, Verona - Springer-Verlag, London, 1987.

Rest in the treatment of respiratory muscle fatigue

P.T. MACKLEM
Meakins-Christie Laboratories, McGill University, Montreal, Quebec, Canada

SUMMARY: Conditions allowing respiratory muscle relaxation in patients needing mechanical respiratory support are examined. The survey includes evaluation of respiratory effort in machine-initiated and patient-initiated modes of mechanical ventilation.

Reference is also made to ongoing studies on the measurement of inspiratory flow rate in pressure-cycled ventilators, rest treatment of the inspiratory muscles in COPD and supplementation of respiratory muscle training with cuirasse treatment at the patient's home.

KEY WORDS: Respiratory muscle fatigue; mechanical ventilation; inspiratory flow rate; pressure-cycled ventilators; rest treatment; inspiratory muscles in COPD; respiratory muscle training; cuirasse; patient's home.

Providing rest for the respiratory muscles, thus allowing recovery from fatigue, is one of the putative benefits of mechanical ventilation[7]. If this concept is valid it would be useful to have a simple means of monitoring the degree of inspiratory muscle relaxation, in order that ventilatory parameters may be selected to achieve this goal.

In the first part of my paper I will present a simple, non-invasive method for accomplishing this[9].

The pressure in an endotracheal tube relative to atmospheric pressure represents the pressure across the respiratory system during positive pressure ventilation. Monitoring this pressure allows identification of relaxation conditions by simple visual inspection. Furthermore, when the patient is contributing to the work of inspiration, measurement of the deviation from the relaxation curve allows quantification of the pressure time product of the respiratory muscles[2] and (along with measurement of lung volume change) the work of breathing[4,5].

In the current study[9] we have defined the conditions under which respiratory muscle relaxation was achieved in a group of patients in whom respiratory muscle weakness prevented discontinuation of mechanical support. The inspiratory muscle pressure-time product and inspiratory work of breathing during both machine initiated (CMV) and patient initiated (AMV) modes of mechanical ventilation were measured and the effects of changes in ventilatory parameters were explored.

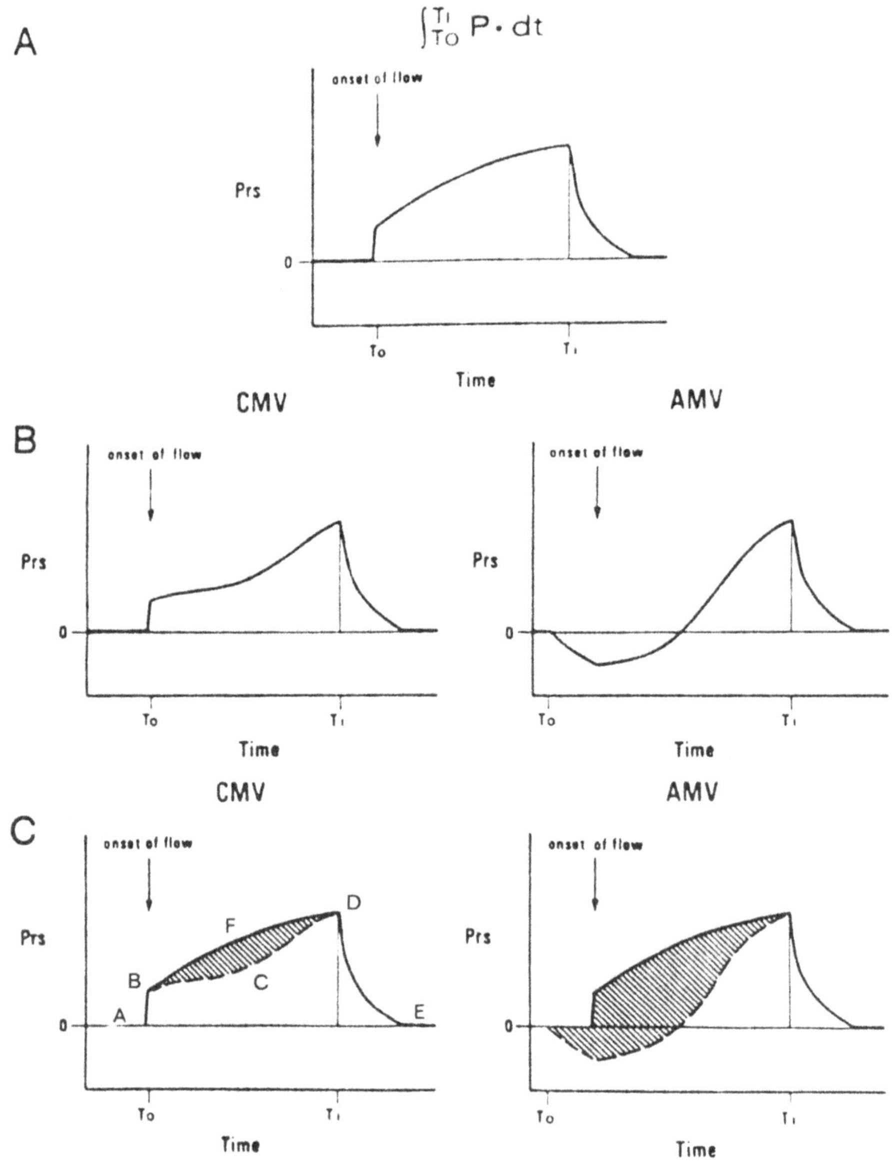

Fig. 1 - Subtraction of the area subtended by the inflation pressure (Prs)-time curve in the presence of inspiratory muscle activity (panel B) from that recorded during passive inflation (panel A) yields the pressure-time product ($\int$P.dt) of the inspiratory muscles.[9]

When a mechanically ventilated patient is completely relaxed, the pressures developed by the ventilator are the pressures required to inflate the relaxed respiratory system, including whatever pressure is required to overcome intrinsic PEEP and the flow resistance of the respiratory system. By simple inspection of the pressure curve during volume cycled mechanical ventilation, one can easily see whether the subject is relaxed or not. During relaxation, the pressure swing never becomes negative during CMV. For any given ventilator setting, the pressure wave is highly reproducible from breath to breath (Fig. 1A), whereas under circumstances when the patient is not conforming well to the ventilator, and is contracting his own inspiratory muscles, there is considerable breath-to-breath variation in the shape of the pressure wave. In this instance, a gradual rise in transrespiratory system pressure is not seen: it may begin to rise, but soon fluctuations occur throughout (Fig. 1B), and if one compares this with EMG of the diaphragm, one can see that, coincident with each dip in transrespiratory system pressure, there is electrical activity of the diaphragm. In the AMV mode, by definition muscles cannot be at rest, because they must trigger the machine in order to start the machine going. Indeed the pattern of muscle recuitment under the assist-control ventilation mode is different than in the control mode, where the diaphragmatic recruitment comes on during midinspiration: in the assist-control mode it must come on at the beginning of inspiration; often, however, it continues considerably after the machine starts to take over, and it is only later during the cycle that the diaphragm becomes relaxed. In fact, the rationale of assist-control ventilation, whereby one makes an inspiratory effort to trigger the machine, and when the machine triggers one immediately relaxes and allows the machine to do all the work, is a very strange rationale: why would anyone ever think that once the inspiratory centers get turned on in the brainstem, an event external to the body would turn them off. The data that Ward and we have obtained at the Montreal Chest Hospital[9] show that the electrical activity, after triggering the ventilator, indeed goes on for a considerable duration during the subsequent inspiration. Now, provided the ventilator settings are identical, the pressure wave can be superimposed on the curve during relaxation as shown in Figure 1C. The area under the relaxation curve gives you the tension-time index of the machine in producing ventilation in a relaxed patient. If the patient were breathing on his own, the area under ABCD and back to the baseline would give you the tension time index of the patient's respiratory muscles breathing spontaneously. Under circumstances when the patient is not conforming well to the machine, the area ABFD and back down to the baseline gives you the tension time index performed by the machine, and the difference between the active inspiration and the passive inspiration (the hatched area BCDF) gives you the tension time index of the patient's own respiratory musculature.

In the assist – control mode, as can be seen in Figure 1 B and C, the situation is quite different: the pressure wave shows that the patient was never relaxed except at the very end of inspiration: he must develop a pressure to trigger the machine, but in fact he has to develop a greater pressure than that in order to overcome intrinsic PEEP, as Milic-Emili has pointed out, before flow actually starts; the hatched area, indicating the TT index for the patient's inspiratory muscles during assist-control mode at this particular inspiratory flow rate, is about the same as du-

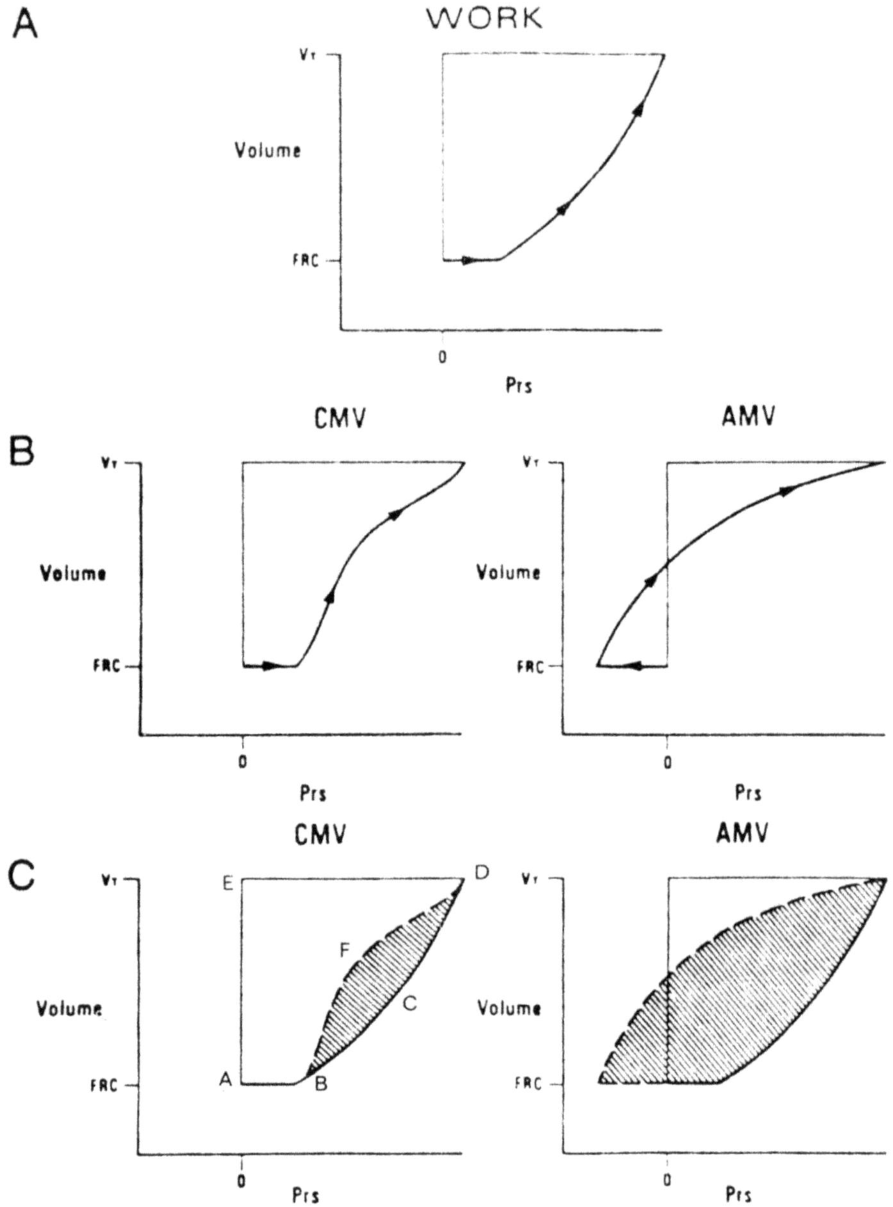

Fig. 2 - Subtraction of the area subtended by the inflation pressure (Prs)-volume curve in the presence of inspiratory muscle activity (panel B) from that recorded during passive inflation (panel A) yields the work of inspiration (shaded area, panel C).[9]

ring relaxation, indicating that, even though the ventilator is functioning, the TT index of this patient's muscles at this ventilator setting is approximately equal to what it would be if the patient were breathing spontaneously on his own.

The work of breathing can be obtained if volume is also measured[1]. Figure 2 shows volume-transrespiratory pressure plots during CMV and AMV; the work performed by the machine when the patient is relaxed is given by ABCDE, when the patient is not relaxed by ABFDE, and the work performed by the patient's own muscles by the hatched area BCDF. Similarly, for the assist-control mode, the work performed by the patient's muscles is given by the hatched area, virtually identical to the work performed if he were breathing entirely on his own.

What are the best ventilator settings to get maximum relaxation?

In all patients we studied, in relaxed and nonrelaxed curves at different inspiratory flow rates during CMV and AMV the hatched area decreased as inspiratory flow increased, therefore the best ventilator settings, if one wishes to rest the patient's respiratory muscles, are obtained at high inspiratory flow rates.

Figure 3 shows the work of breathing, or the TT as a function of inspiratory flow rate for control ventilation and for the assist-control mode (at a ventilator sensitivity of –2 cmH_2O): even at high inspiratory flow rates, the work or the TT index performed by the patient's inspiratory muscles is a substantial amount of the total, and it only approaches zero, under conditions of control CMV, at flow rates in the order of 1 L/sec. Thus, simple monitoring of endotracheal pressure can give valuable information on the work and the TT index the patient is performing.

From this work we have concluded that the best ventilator settings, to get the best degrees of rest are those at flow rates of at least 1 L/s and respiratory frequencies which are as close as possible to the patient's own inherent respiratory frequency.

This method of monitoring the degree of relaxation, as I pointed out earlier, is only useful for volume-cycled ventilators; only when volume is the independent variable can pressure be used as a measure of how much the patient is doing on his own. If you use pressure-cycled or cuirasse ventilators, then pressure is the independent variable and the resulting volume or inspiratory flow rate is the dependent variable, and it is, at least theoretically, possible to measure inspiratory flow rate with pressure-cycled ventilators and do exactly the same analysis as was done with pressure when it was the dependent variable. Studies are currently under way at McGill University to see if it works as well as the measurement of pressure when volume-cycled ventilators are used.

What has been discussed to this point is applicable only to patients in acute respiratory failure, who are intubated. What about patients who are stable, but who are chronically short of breath and unable to exercise and who may or may not be hypercapnic? Would inspiratory muscle rest benefit them in any way? In contrast to the iron lung, there are cuirasse-type ventilators, in which a cage fits over the rib cage and abdomen; the patient gets into a suit, hooked up to a suitable pump, which develops a negative pressure around the body. Patients can be mechanically ventilated at home, at night, at a relatively small cost. This equipment costs less than $ 4000, and is virtually the only cost for this treatment, save for the amount of

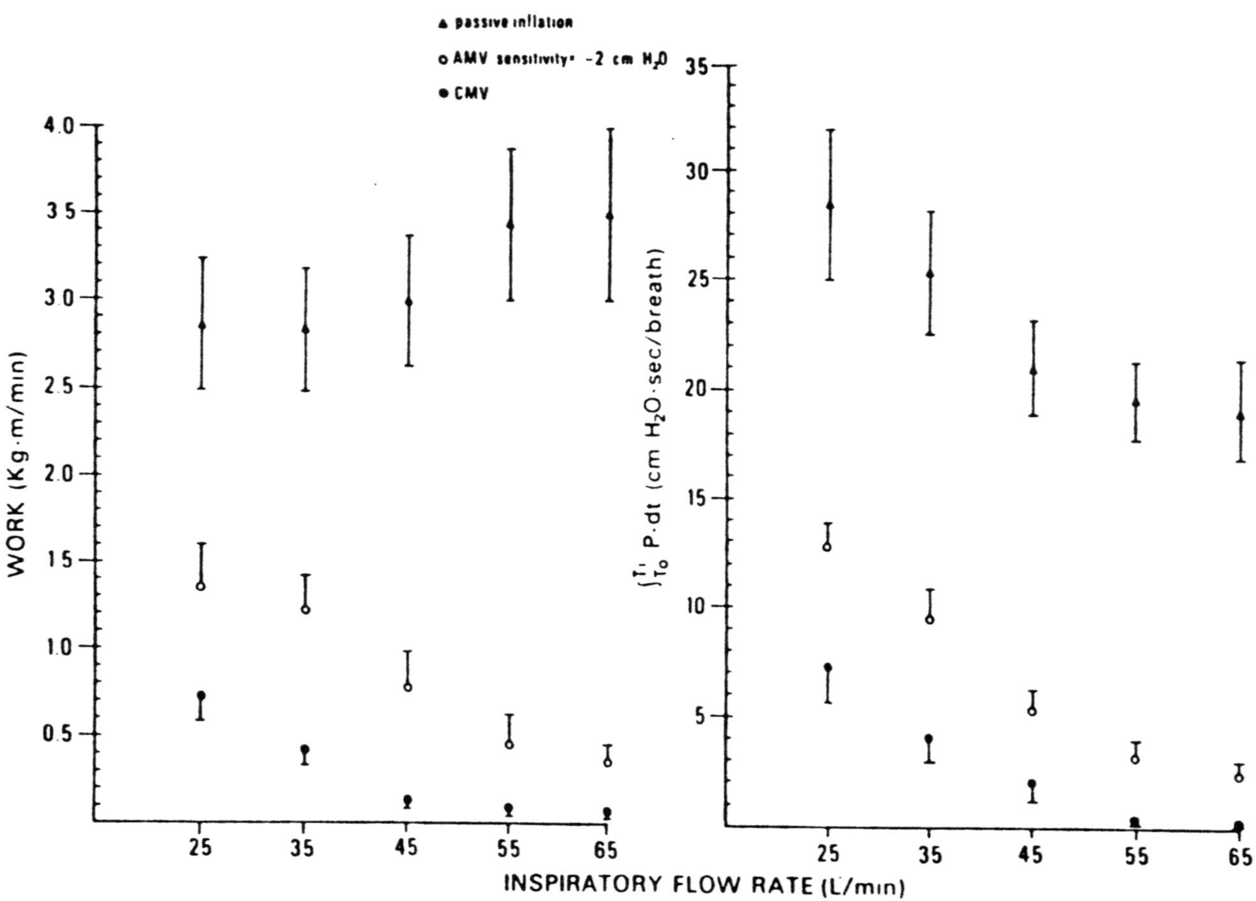

Fig. 3- The influence of inspiratory flow rate on inspiratory work of breathing and inspiratory muscle pressure-time product during CMV (closed circles) and AMV (open circles). The areas under the Prs-volume and Prs-time curves during passive inflation are plotted above for comparison. Values represent mean ± standard error.[9]

electricity required to run the pump. A study, as yet unpublished, was conducted by Pardy at the Montreal Chest Hospital[8] on a number of patients with severe chronic airway obstruction. Thery were given only one hour of rest, using this type of cuirasse ventilator: as we heard, one hour of rest would reverse high frequency fatigue, it should restore maximum strength, it may improve the Edi/Pdi relationship, but should do nothing whatsoever to low frequency fatigue. What was found was: 1) during this period of 1 hour of respiratory muscle rest, there was a fall in total body oxygen consumption, indicating that the inspiratory muscles were using less oxygen than previously; 2) there was a small but significant increase in maximum Pdi after rest, compared to before; 3) in some subjects, there was reversal of the EMG changes of fatigue, i.e. the H/L ratio increased after ventilatory muscle rest, compared to before rest; 4) in some subjects there was a highly significant increase in the Pdi developed for any given Edi. Thus, only 1 hour of ventilatory muscle rest in patients with severe COPD can, in some patients at least, give objective evidence of improvement of diaphragmatic function. This, I think, is important, not only because it shows that diaphragmatic function can be improved, but because as a therapeutic test it also shows that there is such a condition as chronic inspiratory muscle fatigue in which, because the patients must continue to breathe day and night for the rest of their lives, and because their work of breathing is so great, they never get sufficient rest to recover, even though they may be up and about, and not hospitalized. A most important study was conducted by Marino and Braun, still published only in abstract form[3,6], in which 17 patients with COPD, all of whom had hypercapnia, were treated at home with this form of mechanical ventilation used at night. She studied them before and after 2 years of nocturnal mechanical ventilation at home. The results she described are really astonishing; there was no change in FEV, as one would predict, as the underlying disease was not being treated. There was, however, a significant improvement in VC, in maximum inspiratory and expiratory pressures, a reduction in PCO_2 during the day off the machine, from a mean of 54 mmHg to a high normal value of 45, an improvement in the degree of dyspnea and exercise performance, a significant fall in the number of hospital admissions per year from 2.8 to 0.4, and the duration of stay fell from 39 days on average per hospital admission to only 4 days; therefore, the product of hospital admissions per year and the duration of stay decreased the number of hospital days/year from 109 to 1.6. Hospitalization of a critically ill patient with severe COPD and exacerbation in the USA costs in the order of 500 dollars a day or more. With the number of days hospitalized decreased from 109 to less than 2 and a cost of and treatment of less than 4000 dollars, this is going to be, if confirmed, very cost effective.

It is estimated that there are 10,000-20,000 patients in the USA who would benefit from this form of treatment, or approximately 2-3 times the number of patients receiving chronic renal dialysis. It would seem, therefore, that nocturnal ventilator muscle rest at home may be a major new form of treatment of patients with severe

COPD. What is now needed is a randomized controlled clinical trial of this form of treatment, and also of inspiratory muscle training programs, to determine how effective these treatments really are. Such a study is currently under way in Montreal, in which we are studying a control group of patients, who are receiving the usual treatment, a group receiving ventilatory muscle rest, a third group receiving inspiratory muscle training, and a fourth group, receiving a combination of training and rest. The study will take 3 years to finish, so it is too early to give any results, except that we have determined that the problem of obstructive apneas induced by a cuirasse during sleep is not a major problem.

As I stated at the outset of this talk, treatment of the chest wall in order to try and reverse the most crippling aspects of chronic lung disease is still very much a subject of research, and I would caution you that all of the therapeutic measures that you heard about this afternoon are not yet ready for routine clinical use. The only indications, at this stage, for instituting any of these therapeutic measures is as part of a research program to determine their benefit and their cost effectiveness.

References

1. AGOSTONI E., CAMPBELL E.J.M., FREEDMAN S. Energetics. In: CAMPBELL E.J.M. The respiratory muscles: Mechanics and Neural Control. Philadelphia, PA: Saunders, 1970, p 115-42.
2. BELLEMARE F., GRASSINO A. Effect of pressure and timing of contraction on human diaphragm fatigue. J. Appl. Physiol. 1982, 53, 1190-95.
3. BRAUN N.M.T., FAULKNER J., HUGHES R.L., ROUSSOS C., SHAGAL V. When should respiratory muscles be exercised? Chest, 1983; 84, 76-84.
4. MARINI J.J., CAPPS J.S., CULVER B.H. The inspiratory work of breathing during assisted mechanical ventilation. Chest 1984, 87, 613-8.
5. MARINI J.J., RODRIGUEZ R.M., LAMB V. The inspiratory workload of patient initiated mechanical ventilation. Am. Rev. Respir. Dis. 1986, 134, 902-09.
6. MARINO W., N.M.T. BRAUN. Reversal of the sequelae of respiratory muscle fatigue by intermittent mechanical ventilation (abstract). Am. Rev. Respir. Dis. 1982; 125 (pt. 2). 85.
7. MORGANROTH M.L., MORGANROTH J.L., NETT L.M., PETTY T.L. Criteria for weaning from prolonged mechanical ventilation. Arch. Int. Med. 1984, 144, 1012-6.
8. PARDY R.L., MACKLEM P.T., Unpublished observations.
9. WARD M.E., CORBEIL C., GIBBONS W., NEWMAN S., MACKLEM P.T. Optimization of respiratory muscle relaxation during mechanical ventilation. Anesthesiology, submitted for publication.

Key Words Index

www.ingramcontent.com/pod-product-compliance
Ingram Content Group UK Ltd.
Pitfield, Milton Keynes, MK11 3LW, UK
UKHW051450290926
15471UKWH00008B/1115
9781447138525